DEVELOPMENTAL AND CELL BIOLOGY SERIES

EDITORS

P. W. BARLOW D. BRAY P. B. GREEN J. M. W. SLACK

PATTERN FORMATION IN PLANT TISSUES

Development and cell biology series

SERIES EDITORS

Dr P. W. Barlow, *Long Ashton Research Station, Bristol*
Dr D. Bray, *MRC Cell Biophysics Unit, King's College, London*
Dr P. B. Green, *Dept of Biology, Stanford University*
Dr J. M. W. Slack, *ICRF Developmental Biology Unit, University of Oxford*

The aim of the series is to present relatively short critical accounts of areas of development and cell biology where sufficient information has accumulated to allow a considered distillation of the subject. The fine structure of the cells, embryology, morphology, physiology, genetics, biochemistry and biophysics are subjects within the scope of the series. The books are intended to interest and instruct advanced undergraduates and graduate students and to make an important contribution to teaching cell and developmental biology. At the same time, they should be of value to biologists who, while not working directly in the area of a particular volume's subject matter, wish to keep abreast of developments relevant to their particular interests.

BOOKS IN THE SERIES

R. Maksymowych *Analysis of leaf development*
L. Roberts *Cytodifferentiation in plants: xylogenesis as a model system*
P. Sengel *Morphogenesis of skin*
A. McLaren *Mammalian chimaeras*
E. Roosen-Runge *The process of spermatogenesis in animals*
F. D'Amato *Nuclear cytology in relation to development*
P. Nieuwkoop & L. Sutasurya *Primordial germ cells in the chordates*
J. Vasiliev & I. Gelfand *Neoplastic and normal cells in culture*
R. Chaleff *Genetics of higher plants*
P. Nieuwkoop & L. Sutasurya *Primordial germ cells in the invertebrates*
K. Sauer *The biology of Physarum*
N. Le Douarin *The neural crest*
J. M. W. Slack *From egg to embryo: determinative events in early development*
M. H. Kaufman *Early mammalian development: parthenogenetic studies*
V. Y. Brodsky & I. V. Uryvaeva *Genome multiplication in growth and development*
P. Nieuwkoop, A. G. Johnen & B. Albers *The epigenetic nature of early chordate development*
V. Raghavan *Embryogenesis in angiosperms: a developmental and experimental study*
C. J. Epstein *The consequences of chromosome imbalance: principles, mechanisms, and models*
L. Saxen *Organogenesis of the kidney*
V. Raghavan *Developmental biology of fern gametophytes*
B. John *Meiosis*
R. Maksymowych *Analysis of growth and development of Xanthium*
J. Bard *Morphogenesis: the cellular and molecular processes of developmental anatomy*
R. Wall *This side up: spatial determination in the early development of animals*

PATTERN FORMATION IN PLANT TISSUES

TSVI SACHS
Department of Botany, The Hebrew University
Jerusalem, Israel

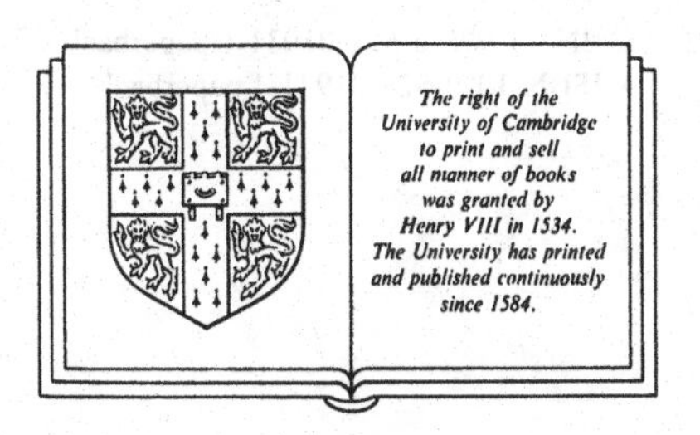

CAMBRIDGE UNIVERSITY PRESS
Cambridge
New York Port Chester
Melbourne Sydney

CAMBRIDGE UNIVERSITY PRESS
Cambridge, New York, Melbourne, Madrid, Cape Town, Singapore, São Paulo

Cambridge University Press
The Edinburgh Building, Cambridge CB2 2RU, UK

Published in the United States of America by Cambridge University Press, New York

www.cambridge.org
Information on this title: www.cambridge.org/9780521248655

First published 1991
This digitally printed first paperback version 2005

A catalogue record for this publication is available from the British Library

Library of Congress Cataloguing in Publication data

Sachs, Tsvi.
Pattern formation in plant tissues/Tsvi Sachs.
p. cm. — (Development and cell biology series)
Includes bibliographical references.
ISBN 0 521 24865 5
1. Plant cells and tissues. I. Title. II. Series.
QK725.S22 1990
581.8′2 — dc20 90-31057 CIP

ISBN-13 978-0-521-24865-5 hardback
ISBN-10 0-521-24865-5 hardback

ISBN-13 978-0-521-01931-6 paperback
ISBN-10 0-521-01931-1 paperback

For Laura

Contents

Preface

The central topic and justification of this book is a conceptual problem: how could orderly or patterned biological form be based on chemical and physical processes. Form is known to be the product of development – but development can lead to tumors rather than organized tissues. Form is also known to depend on the genetic constitution of the organism. Yet this is only a restatement of the problem, since any given gene specifies the structure of a molecule, not the organization of tissues.

The use of new molecular methods is sure to yield a wealth of information. The search for this information requires a conceptual framework of how tissue patterns are formed. Yet many accepted views are implied rather than clearly stated. Furthermore, they consist of unquestioned dogma which is partially contradicted by available facts. An alternative set of concepts is sought here on the basis of a broad, comparative view of available facts, mostly the products of simple techniques. Such a framework should also be important for studies of developmental physiology at non-molecular levels. Finally, though it is the genes that mutate and it is mature structures that are screened by selection, the intervening controls of orderly development must impose constraints on the possibilities exposed to selection – and thus on the course of the evolution of biological form.

Though the central question is theoretical, it is assumed here that at the present state of knowledge theory devoid of observations and experiments is not likely to be useful. Only facts can constrain the almost unlimited theoretical possibilities. A factual approach requires the discussion of defined developmental events and experimental systems. Yet the conceptual problem is the central purpose here, so facts are mentioned only where they are deemed to contribute to theoretical discussion and references to earlier work are meant to be useful rather than historical or laudatory. This book is therefore primarily a series of related research essays, each chapter dealing with a defined problem and meant to be as self-contained as possible. These chapters are followed by a discussion which is meant to develop an alternative or at least supplementary approach to the problem of the specification of form.

The discussions are limited to plants. This is a relatively advantageous group for the consideration of the development of biological form. Plants share most aspects of cell biology with other organisms and past experience has shown that principles discovered in plants can be of general biological importance. The book is therefore an attempt to continue and update the work on internal controls of plant development of many early authors, especially Vöchting (1892), Jost (1907), Bünning (1953), Sinnott (1960), Steward (1968), Wardlaw (1968) and Steeves & Sussex (1972). It differs from the more recent volume edited by Barlow & Carr (1984) in dealing with many different topics from one point of view.

The conclusions and the organization of the material presented here were products of collaborations and discussions with many people. Some of these have also taken the trouble of commenting on early versions of some chapters. Special thanks are due to P. W. Barlow, D. Cohen, J. Croxdale, A. Fahn, M. Gersani, P. B. Green, M. Kagan, B. Leshem, F. Meins, Jr., G. J. Mitchison, A. Novoplansky, L. W. Sachs, A. R. Sheldrake, and E. Werker. Thanks are also due to Y. Gamborg for help with the photographs.

1

Introduction

THE PROBLEM OF PATTERN SPECIFICATION

The form of living organisms is one of their outstanding characteristics: though organisms are complex, their various parts bear predictable, repeated relations to one another. It is this regularity, or the deviation from a random distribution of the various parts, that will be referred to as a *patterned structure* (Child, 1941; Spemann, 1938; Sinnott, 1963; Wolpert, 1971). The regularity of form is characteristic not only of external shape but also of the microscopic arrangements of the various types of cells and tissues (Fig. 1.1). Furthermore, patterns are apparent in the temporal as well as the spatial relations. The following chapters are concerned with the processes that generate cellular patterns in plant tissues, with the basis of organized, in contrast to tumorous, structure.

Mature biological form is necessarily the result of the non-random distribution of the developmental processes. Form can be changed and even disrupted without preventing the continuation of growth and differentiation: examples are the regenerative events following wounding, the development of callus and tumor tissues and the effects of various mutations. It follows that it is possible to separate the 'building blocks' of development, the processes of growth and differentiation, from the regulation of their spatial and temporal occurrence (Wolpert, 1971). Thus discussions of patterning can often accept as given the molecular and cellular processes of growth and differentiation, fascinating though they are, and deal primarily with the determination of their relative location, in both time and space.

Since form is predictable, it must be specified in some way. This specification is present even in single cells, in zygotes or in somatic cells that regenerate whole plants in tissue culture. Even though environmental signals modify plant form, and these modifications are larger and more common than comparable modifications of animal development, internal specification must have a dominant role. The role of the environment is necessarily limited – plants develop their characteristic form without environmental clues and even in varied conditions. It is perhaps more

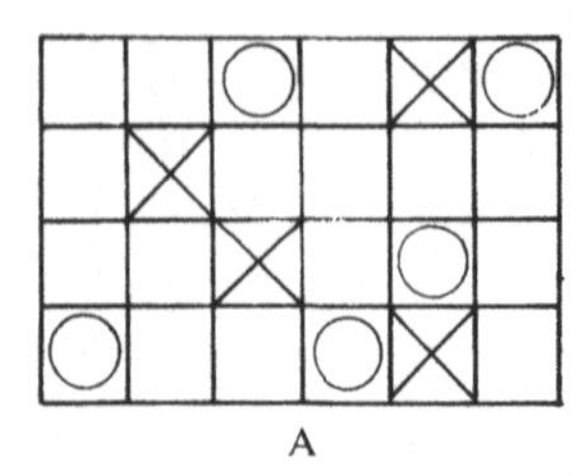

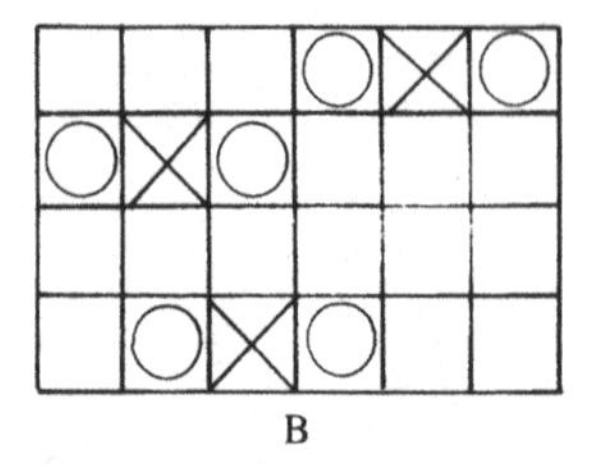

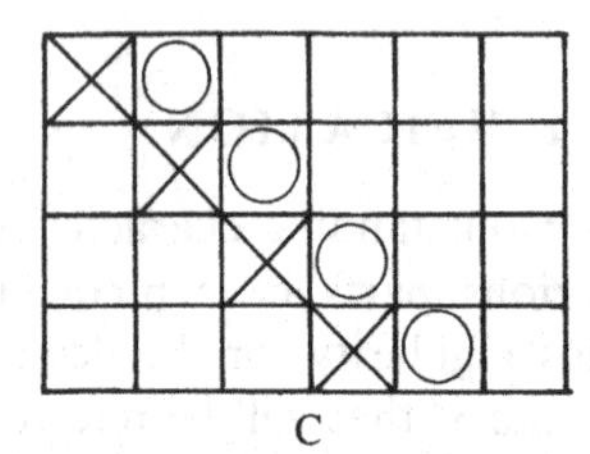

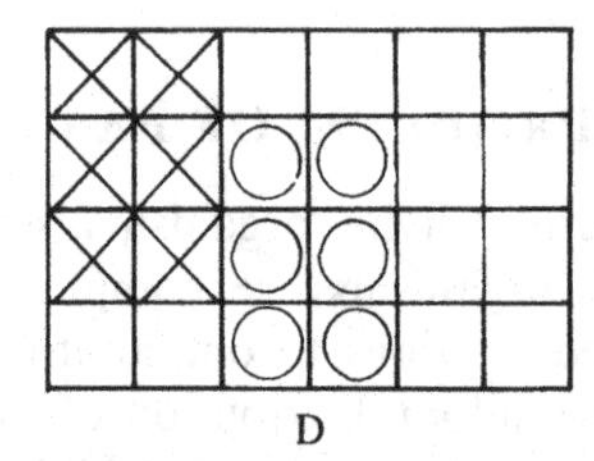

Figure 1.1 Schematic representation of cellular patterning. Two types of specialized cells are distributed in a matrix of a third cellular type. A. Judging from a limited field, the distribution of the specialized cells could be random. B–D. The distribution of the specialized cells is not random, since the location of any given cell is at least partially predictable from the location of other cells. The mechanisms responsible for the development of these and additional cellular patterns are the subject of this book.

significant that differences in environmental conditions are too coarse, do not have a fine enough grain, to specify the details of the location of individual branches, and the environment could certainly not determine the precise location of various cell types.

It has been known since Mendel that the genes specify a part of, if not all, the details of biological form. In view of present knowledge of the nature of genetic information, a specification of form by genes only raises the original problem: How the structure of macromolecules, nucleic acids and proteins, could be causally related to the pattern not of crystals but of cells, tissues and even large organs. Thus though genes specify structures, they can only do this as parts of a larger, coordinated developmental system. The action of any gene, even if isolated and fully sequenced, could only understood as part of broader, integrated 'whole'. Biological patterning is a major expression of this integration.

MAJOR CONCEPTS OF BIOLOGICAL PATTERNING

Orderly or patterned development, and its capacity for regeneration following severe damage, have been the basis of many 'vitalist' ideas

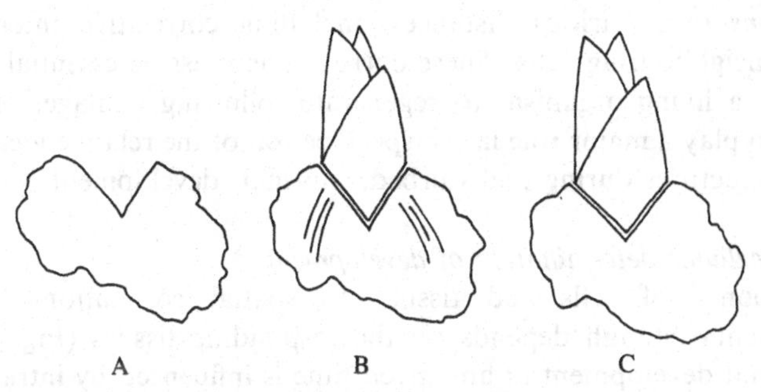

Figure 1.2. A simple experimental demonstration of both an induction of differentiation and of tissue determination. A. Control: callus grown in culture was wounded. This wound did not result in any differentiation. B. A bud was grafted into the wounded callus. The presence of the bud was correlated with the differentiation of vascular cells in the callus. It follows that the bud was a source of signals that *induced* specialized differentiation in the callus. C. The treatment was the same as in B, but the callus did not respond by any differentiation. The cells of this callus are said to be *determined*, or not to be *competent* to respond to the signals of the grafted bud (based on Camus, 1949).

(Gould, 1977) and of claims that there is no conceivable way in which the chemical and physical structure of a cell, such as a zygote, could specify its own development into a complex multicellular organism. The idea of an homunculus was a primitive answer to this dilemma; it suggested that the zygote included a precise blueprint of the mature organism – and of all its future offspring. A more sophisticated variation of this blueprint approach is a strict molecular program in which each event necessarily leads to the next, more complicated stage. These concepts do not account for the regenerative capacities of most living organisms: neither a blueprint nor a program could be responsible for the ability of developing organisms to complete themselves after unpredictable damage. At present, however, there is no lack of quantitative models that demonstrate the possibility of a molecular specification of development (Brenner et al., 1981; Meinhardt, 1982; Malacinski, 1984). These plausible models use the following major concepts.

(a) The induction of differentiation

The removal of one part of an organism changes the developmental fate of the remaining parts. Furthermore, grafting embryonic tissues in unusual locations can change the fate, or *induce* the differentiation, of neighboring regions (Fig. 1.2; Spemann, 1938). These general facts, demonstrated for plants about a hundred years ago (Vöchting, 1982; Jost, 1893), show that the developmental fate of a cell or tissue depends on its location: it can be influenced by signals from other parts of the organism (Bünning, 1953, 1965; Lang, 1973). It follows that development could involve *spatial correlations* over various distances, including correlative interactions between neighbouring cells. These correlations must be essential for the ability of a living organism to regenerate following damage, but they could also play a major role in the specification of the relative location of various structures during undisturbed, 'normal' development.

(b) *Intracellular determination of development*
The response of cells and tissues to spatial correlations and to environmental stimuli depends on the responding tissues (Fig. 1.2). It follows that development at any given time is influenced by intracellular factors, ones that are a function of the developmental history of the cells and not only on the conditions they are in at any given time (Wareing, 1978; Meins, 1986). The role of these intracellular factors can be expressed in three ways. Developmental events may be determined, as shown by their stability even when conditions change. A second expression of determination includes not only a stability of a state but also of a continued developmental program, in which one event necessarily leads to another, regardless of spatial signals. Finally, determination may be expressed by different responses to the very same signals: the cells differ in their *competence* to respond to different conditions (Osborne & McMannus, 1986). These various expressions of the dependence on past developmental events, or *temporal correlations* (Sachs, 1978a), are a source of the stability characteristic of developmental processes.

(c) *The regulation of gene expression*
Present studies of the processes of differentiation accept that most cells have the entire complement of genetic information. Almost all differentiation thus consists of the selective expression of different parts of this information, not of an unequal distribution of the genetic material itself (Hertwig, 1890). The expression of various genes is correlated with one another, and there can be homeotic or master genes whose expression leads to major, far-reaching effects. The processes of differentiation itself are not the topic here, but concepts of gene regulation are essential because they allow for concrete ideas of *controls of development*. Thus, gene regulation shows how low concentrations of simple substances could elicit large developmental responses, responses that are not related in any obvious way to the chemical structure of the signals.

(d) *Pre-patterns*
The possibility of gene regulation suggests that differences between neighboring cells may depend on small differences in the conditions the cells are in at critical stages of their development (Fig. 1.3). In other words, much of the complexity of development and of the mature structures could depend not on large differences in the specification of cell

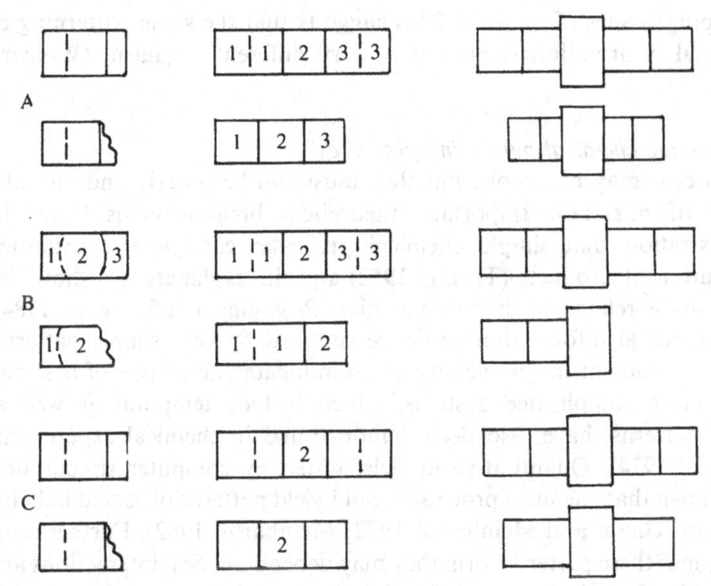

Figure 1.3. Three models of controls of patterning and of pattern regeneration following damage. In all cases the 'organism' has five cells, the central one being specialized or different from the rest. A. A gradient of a substance within the original or mother cells specifies differences between the 5 daughter cells. These differences are observable only in the special differentiation of the central cell. When the system is damaged at an early stage of its development the gradient is regenerated, becoming steeper in the shortened organism. A central region is still specified, though it is shorter and less precisely defined than in the undamaged organism. B. An intracellular developmental program results in the formation of a series of cells that differ from one another. These differences are apparent in the differentiation of the central cell. This pattern does not regenerate when parts of the organism are removed during development. C. Developmental interactions occur only after the cells are formed; there is no pre-pattern in the mother cell. The special location of the central cells – they are the ones that receive signals from both sides – leads to their special differentiation. These patterning processes occur during rather than before development. Such patterning supports regeneration when the immature organism is damaged.

fate but rather on the cellular expression of this specification (Child, 1941; Stern, 1968; Wolpert, 1969, 1971, 1981). This means that there could be *pre-patterns* or *positional information*, in the form of gradients or other patterned distributions of simple molecules, and these could be much simpler than the resulting, mature patterns.

Pre-patterns could undergo regeneration (Fig. 1.3A) and they suggest chemical or physical meanings for the vague concept of developmental fields (Weiss, 1950; Waddington, 1966; Wardlaw, 1968). Furthermore, grafts of animal tissues into hosts that are distant, in an evolutionary sense, show that the location in one organism can be interpreted by the developing tissues of another. This suggests that the same patterning or positional information is present in very different organism (Wolpert, 1981).

(e) *Dynamic chemical patterning processes*
Pre-patterns may be simple, but they must still be orderly and virtually devoid of mistakes. Important theoretical breakthroughs were the demonstration that simple chemical processes can yield a patterned distribution of products (Turing, 1952) and the explanation of how this patterning is related to thermodynamics (Prigogine and Stengers, 1984). Developmental information could be supplied by very simple patterns, such as a gradient in the vicinity of an autocatalytic source of a signal. Much more complicated systems, which include temporal as well as spatial patterns, have also been demonstrated in chemical experiments (Winfree, 1974). Quantitative models, tested by computer simulations, have shown that chemical processes could yield patterns observed in living organism (Gierer and Meinhardt, 1972; Meinhardt, 1982). Furthermore, at the times these patterns form they may depend on local interactions and be capable of regulation, even after drastic changes. Such regulation could be a basis for biological regeneration.

These concepts are very general. The way they could be applied to real systems allows for many variations and the differences could be quite significant (Fig. 1.3). The important conclusion is that there is no theoretical problem in accounting for patterning on the basis of known and even simple chemical and physical principles. The complexity of biological patterns poses no conceptual problem because it could result from the interaction of two or more simple mechanisms. Furthermore, biological form is certainly specified in stages, each depending on stable or determined states reached during previous development. Thus at present there would be no justification for Waddington's (1966) statement concerning the development and regeneration of *Micrasterias*, that 'we just have to realize we don't have an answer that makes sense'

THE PURPOSE OF THE BOOK

The concepts outlined above relate to the general question how patterns could be specified, not to how biological patterns actually develop. When real cases are considered no simple picture is as yet possible; there is a bewildering array of results and suggestions based on different organisms or biological systems (Malacinski, 1984). Though available hypotheses need not be wrong, there may be need for modifications and even for additional concepts. Plant structure and development offer at least two general indications that a specification of development by chemical pre-patterns or blue-prints may not suffice. The first is that both mature

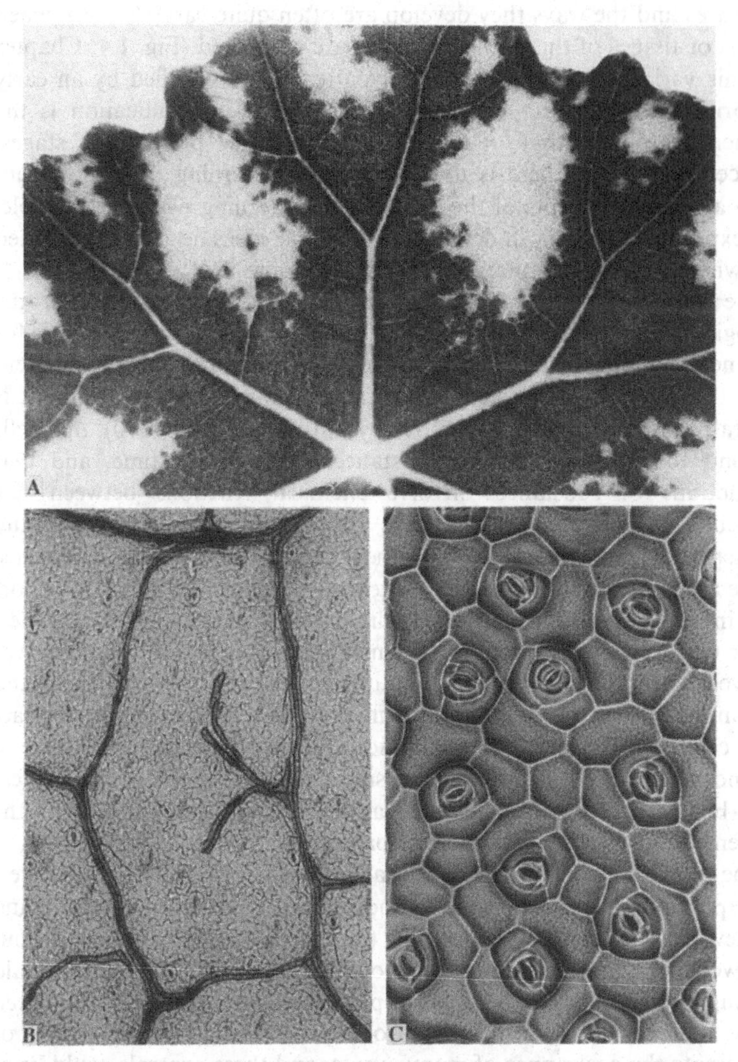

Figure 1.4. Expressions of order and chance in cellular patterns. A. Regions with and without anthocyanins in a *Begonia* leaf. The general pattern of pigmentation is predictable, or is repeated in accordance with the genotype and the environment. However, there is no precise border between pigmented and non-pigmented regions. The determination of cell fate is expressed only by the probability that a given cell will or will not be pigmented. B. Minor veins in a pea leaf. The vein system – what is actually seen is the xylem – reaches all parts of the leaf, but its cellular details are highly variable. C. Stomata in the epidermis of a *Begonia* leaf. These stomata form a definite pattern, yet the details of the distances between neighboring stomata vary greatly. Additional variability is seen in the precise cellular configuration surrounding each stoma. X 2, 13, and 100, respectively.

structures and the ways they develop are often quite variable, even when organs or tissues of the very same plant are compared (Fig. 1.4; Chapter 7). This variability is not expected if patterning is specified by an early blueprint or by a strict program. A second, related indication is the regenerative powers that are not restricted to early, 'pre-pattern' stages. The central purpose here is to consider the patterning of plant tissues using available concepts of the controls of patterning wherever possible. It is expected that this will define how available ideas need to be modified and where additional concepts become necessary.

The need to modify available models could be the result of the biological characteristics of systems being patterned, which allow for parameters of rate, distance and complexity that are dramatically different from those found in simple physical and chemical systems. Thus, it is generally assumed that signals move by diffusion (Crick, 1970), that cells respond to concentrations of substances at a critical time, and that interactions of substances involve chemical reactions between few molecules. These assumptions have the advantage of simplicity and mathematical modelling shows how they could result in beautiful patterns. These assumptions, however, are quite arbitrary: there is active transport in living cells and diffusion may be limited by impermeable membranes. There can be virtually unlimited chemical reactions in a cell that receives a developmental signal, so the quantitative relations between signal and response need not reflect simple chemistry. As a final example, the fact that cells can undergo determination means that they can have a 'memory' and need not respond to signal concentrations at one critical time but rather to more reliable parameters, such as changes of the concentration of critical substances or signals.

The approach followed here is that not only new facts but also new concepts could be required so as to understand the generation of form and the newly discovered facts concerning the genetic controls of development. The working hypothesis is that the modification of current models should be sought in biological parameters of patterning processes. It is considered likely that there are central controls, processes which determine or coordinate the occurrence of many events, and these controls could be at a cell or even a tissue level. It is thus necessary to investigate specific though major examples of patterning while searching for a broader picture. The general aim is not to understand molecular mechanisms, desirable though these would be, but rather to define general principles concerning the specification of biological form. The approach is reductionist concerning individual phenomena but the aim is to define missing 'holistic' or integrating properties of developmental systems.

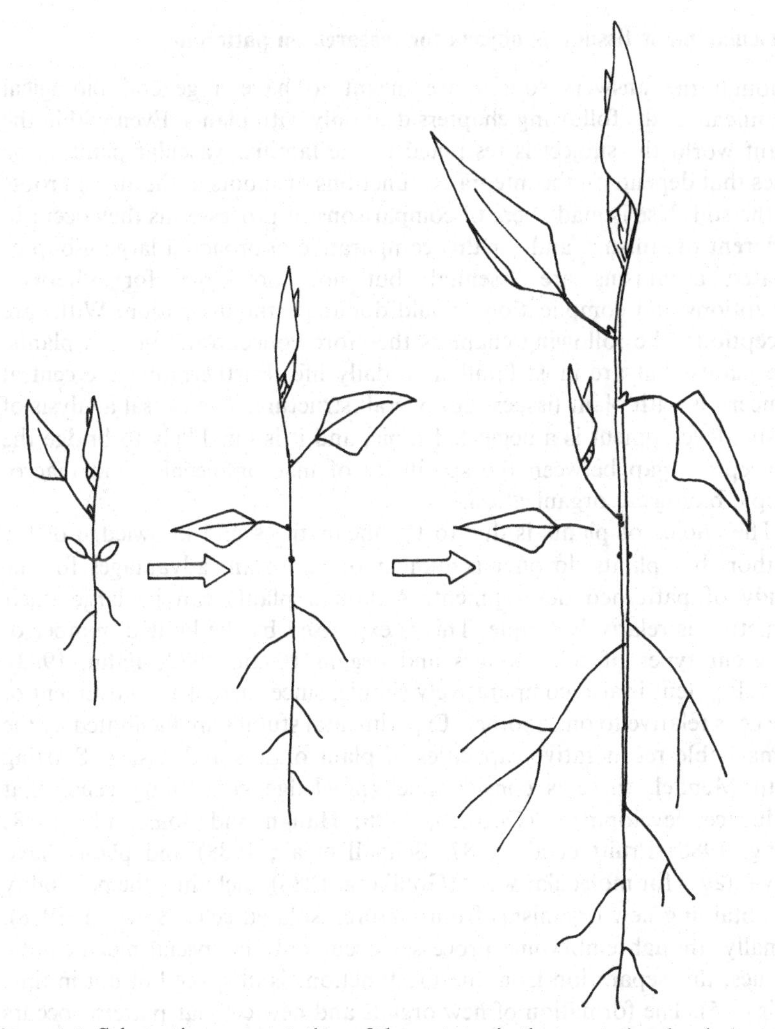

Figure 1.5. Schematic representation of three stages in the vegetative development of a vascular plant. Both growth and the formation of new organs continues throughout the life of the plant. This growth is restricted to the apical meristems, at the tips of the shoots and roots. Additional apical meristems are formed in the axils of leaves and close to root tips. In most plants there is also growth in girth, adding vascular and supporting tissues that connect the new leaves and the new roots and replacing old tissues. This growth, which involves the formation of new patterned tissue but not new organs, occurs in a meristem called the vascular *cambium*. The tissues that preform the various non-developmental functions of the plant do not grow and have a limited life span. This is most clearly seen in the shedding of older leaves.

Vascular plant tissues as objects for research on patterning

Though the answers sought are meant to have a general biological significance, the following chapters deal only with plants. Even within the plant world the subject is restricted to the familiar vascular plants – the ones that depend on the integrated functions of shoots in the air and roots in the soil. Use is made here of comparisons of processes as they occur in different organisms, and for this comparative approach a large group of related organisms are essential – but not too large, for otherwise exceptions and complications would dominate the discussion. With rare exceptions, the following chapters therefore concentrate on seed plants, the plants that are most familiar in daily life. Furthermore, the central concern is with plant tissues, not overall structure. The causal analysis of tissue development is a neglected topic, and it is one likely to bridge the conceptual gap between the specificity of macromolecules and macroscopic biological organization.

The choice of plants is due to the inclinations and knowledge of the author, but plants do offer a number of important advantages for the study of patterned development. Although plants can be large, their structure is relatively simple. This is expressed by the limited number of different types of cells, tissues and organs (Esau, 1977; Fahn, 1982). Development is also comparatively simple, since there is no movement of the cells relative to one another. Experimental studies are facilitated by the remarkable regenerative capacities of plant organs and tissues. Starting with Mendel, there is considerable knowledge concerning genes that influence development (Gottlieb, 1986; Haugn and Somerville, 1988; King, 1988; Pruitt et al., 1987; Schnall et al., 1988) and plants have advantages for molecular work (Goldberg, 1988), including the possibility of obtaining new organisms from mature, isolated cells (Steward, 1968). Finally, though embryonic processes occur only in special meristematic tissues, this separation from mature functions is in space but not in time (Fig. 1.5). The formation of new organs and new cellular patterns occurs throughout the life of plants, even when they are thousands of years old, so descriptive and experimental work can be carried out on vigorous seedlings and, when appropriate, even on large trees.

2

Interactions of developing organs

Removing the shoot of a pea seedling results in rapid growth, obvious within two days, of one or more buds (Thimann & Skoog, 1933; Fig. 2.1). The growth of the same buds on intact plants is not noticeable without magnification. This simple experiment is a demonstration in a defined system of a common phenomenon: whenever plants are pruned or eaten by animals their growth is modified so that the missing parts are replaced, leading to a partial or even complete restoration of the original form. This common result must mean that as long as the apical parts of the shoot are present they inhibit, or dominate, the growth of the lateral buds. This relation between the growth of shoots on the same plant has been called 'apical dominance' and has been the subject of many studies (Philips, 1975; Rubinstein and Nagao, 1976; Thimann, 1977; Hillman, 1984).

The significance of apical dominance to the control of form is that it is a simple demonstration of a *developmental correlation* between the parts of a plant (McCallum, 1905). As will be seen below, these macroscopic correlative relations influence not only the development of organs but also the structure of tissues, the main subject of this book. The experiment illustrated in Figure 2.1 raises the questions of the nature of the signals that pass between the dominant shoot and the buds and of the mechanisms by which these signals exert their effect. These questions have been studied (Thimann, 1937, 1977; Rubinstein and Nagao, 1976; Hillman, 1984) but they are considered only in the following chapter. Here a comparative approach is followed: a general survey is made of the various effects of shoot apices and the corresponding correlative effects of other organs (Söding, 1952; Dostál, 1967; Sachs, 1986). Such comparisons have been largely neglected by physiologists. An attempt will be made here to show that they yield important conclusions concerning interactions of plant organs and the ways in which various correlative effects could be integrated. A comparative approach also yields a better basis for a discussion of the signals responsible for these correlations.

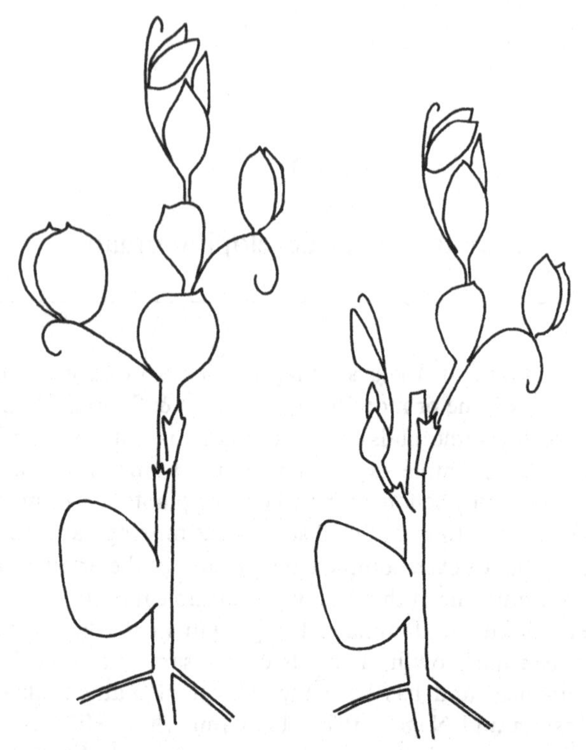

Figure 2.1. Apical dominance in pea seedlings. On the left is an intact plant; on the right the upper part of the shoot was removed. Four days later the buds on the decapitated plant had started growing while the growth on the intact plant was not obvious without magnification. These plants represent a result which occurs with no exceptions and is common to many if not most plant species.

GENERALIZATIONS CONCERNING RELATIONS BETWEEN ORGANS

Scientific literature and agricultural practice provide many facts concerning developmental changes that occur when different organs are removed. The meaning of these for correlative interactions is corroborated by the 'addition' of organs, by grafts or during adventitious regeneration, in unusual, even unpredictable, locations. These 'additional' organs leave no doubt that correlations between the presence of an organ and developmental events are not due to chance. The array of available facts is very large: developmental responses vary depending on the species, age of the plant and environmental conditions. Although this variation is great, some broad generalizations (Sachs, 1986, 1988c) are still valid, since they are applicable to numerous cases and are not contradicted by available facts. This statement does not mean that a generalization about a developmental correlation is always apparent. Any developmental event depends on many factors; when one of these is limiting, changes of other factors, such as inhibitory or inductive signals, need not have any developmental expression (Thimann et al., 1971). The following four generalizations cover processes which occur in most plants.

(a) A growing organ inhibits the development of similar organs

The most common expression of this inhibition is the one shown in Figure 2.1: the growth rate of lateral buds is low as long as the shoot above them is present. The extreme form of this correlation is seen when the growth of the lateral buds on the intact plant is not noticeable. In most species, however, a number of shoot apices grow together, on the same plant. The growth of these apices is still correlated though there is no absolute dominance: when rapidly growing apices are removed the growth rate of the remaining apices is increased (Fig. 2.2A; Jacobs et al., 1959). There are therefore quantitative variations in the degree of apical dominance (Söding, 1952; Sachs and Thimann, 1967). The same principle of inhibition by growing shoots is also expressed in the initiation of new adventitious apices (Fig. 2.2D), apices that form in unusual locations by a re-differentiation of cells. It is the removal of shoots, and not the wound that must accompany the removal, that results in changed growth rates in the rest of the plant. This is shown, for example, by the growth of buds when the development of the main shoot is stopped by encasement in gypsum (McCallum, 1905).

Inhibition of similar development is also exerted by other plant organs. Leaf growth increases when leaves are removed (Arney, 1956). It is general agricultural experience that flower and fruit removal promotes the formation and development of other flowers and fruits on the same plant. The same is also true for the root system, as is apparent from the formation of new roots on cuttings from which the entire root system has been removed. Apical dominance of roots has also been demonstrated when only the main growing root tip was removed (Fig. 2.3A; Torrey, 1950; Wightman and Thimann, 1980). As a general rule, however, the developmental correlations involving roots are less obvious than in the shoots. There appear to be two reasons for this difference. Roots of intact plants branch much more than shoots, so that changes in lateral roots are difficult to detect. A second reason is that correlative influences on root development are expressed in the initiation and primordial growth of apices and not in the expansion growth that is more readily measured (Keeble et al. 1930; Chapter 10). Thus apical dominance in roots is most clearly expressed when the shoot is removed and root initiation in the presence of the dominant root apices is drastically reduced (Fig. 2.3B).

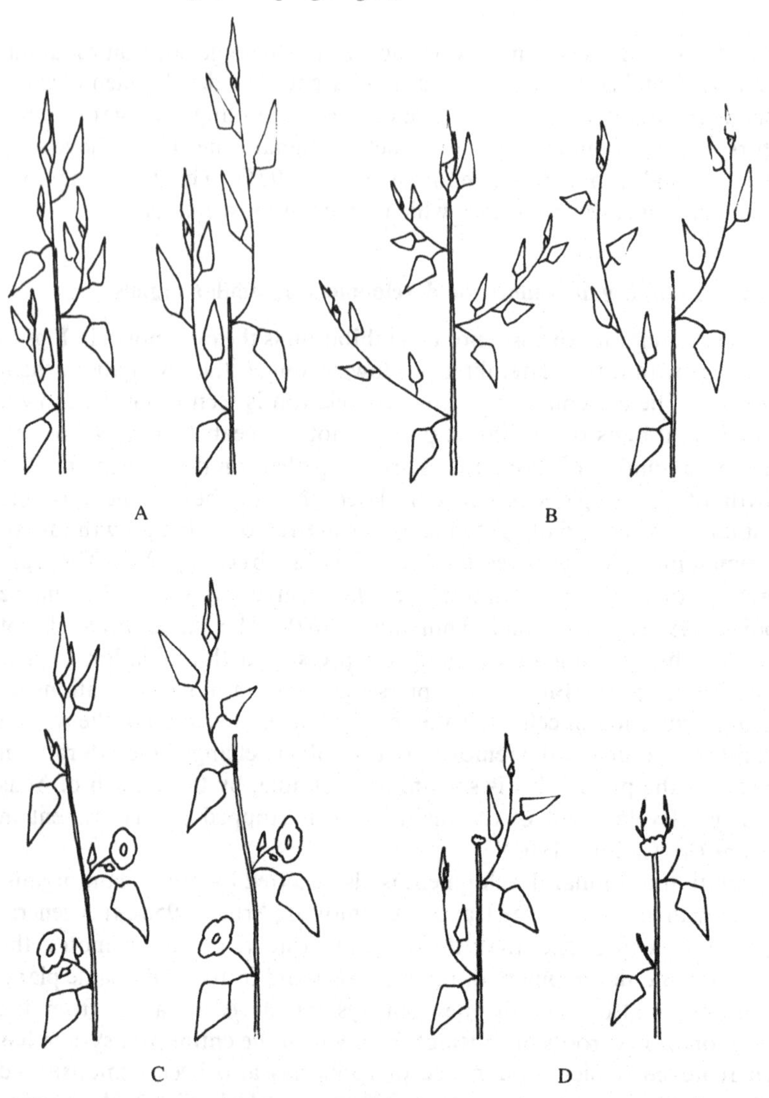

Figure 2.2. Diagramatic representation of some variations in the expression of apical dominance. In every case an intact plant is on the left and a plant from which the top of the rapidly growing shoot was removed is on the right. A. Lateral buds grow even on the intact plant. This growth, including the formation of new leaves, is much more rapid if the main shoot apex is removed. B. The lateral buds grow, but they assume a characteristic, horizontal orientation. These laterals turn upwards when the main shoot apex is removed. C. The lateral buds grow as special, differentiated shoots – in the illustration they carry flowers. When the main shoot is removed, buds that have not yet been committed to special differentiation become dominant, vegetative shoots. D. The removal of all shoot apices, including those of the lateral buds, leads to pronounced callus growth. In some plants there is also a formation of new buds, in unusual or 'adventitious' locations.

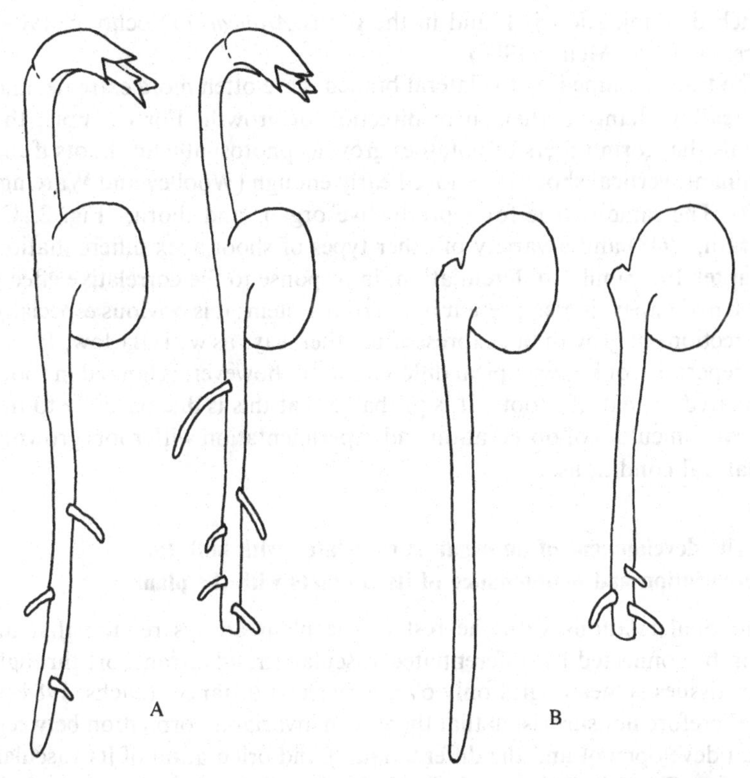

Figure 2.3. Apical dominance in roots of pea seedlings. The tip of the main root was either intact (left) or removed (right). A. Removal from an otherwise intact seedling was associated with increased lateral root formation, but since lateral roots developed even on the controls with an intact main root, the effect was difficult to detect. B. The entire shoot system, with all buds, was removed from both seedlings. The intact root (left) continued to grow, at the expense of the cotyledons, at a rate comparable to that of intact seedlings – but lateral root formation was greatly reduced. Roots on plants with no shoots branched when the apex of the main root was removed. This is a demonstration of apical dominance in roots, a dominance which is not always expressed.

(b) Potentially similar organs differentiate in the presence of a dominant organ

In many plants the lateral shoots grow, even in the presence of a main shoot, but their growth is plagiogeotropic, horizontal rather than vertical (Fig. 2.2B). When the vertical shoot is removed, it is replaced by a change in the orientation of one or more of the lateral branches (Snow, 1945). The orientation of the laterals can also express a determined differentiation of the apices, a differentiation which does not change even when the main shoot is removed. An extreme and well-known example of such lateral branch determination is found in the genus *Araucaria* (Vöchting, 1904; Wareing, 1978; Meins, 1986).

The traits assumed by the lateral branches are often more extreme, and less readily changed, than mere direction of growth. For example, the laterals that form tubers of potatoes grow as photosynthetic shoots if the dominant, vertical shoot is removed early enough (Woolley and Wareing, 1972). The same is true for reproductive organs and thorns (Fig. 2.2C; Umrath, 1948) and a variety of other types of shoot apex differentiation (Chapter 10). Similar differentiation, in response to the correlative effects of the main apex, is also present in the root system; it is obvious especially in directions of growth but expressed in other ways as well (Barlow, 1986). The repertoire of known apical differentiation, however, is limited in roots compared to that of shoots. It is probable that this is due partially to the relative difficulties of observation and experimentation with roots growing in natural conditions.

(c) The development of an organ is correlated with both the differentiation and maintenance of its contacts with the plant

Functional relations with the rest of the plant always require that an organ be connected by differentiated vascular strands: transport through other tissues is meaningful only over very short distances (Sachs, 1984a). It is therefore not surprising that there is an invariable correlation between organ development and the differentiation and orientation of its vascular contacts (Fig. 2.4). This correlation holds for all shoots and roots and all aspects of vascular tissues: primary, secondary and regenerative differentiation of the phloem, the xylem and of the cambium between them (Sachs, 1981a). In plants with a vascular cambium, and these are the majority, the inductive influence on vascular differentiation continues as long as an organ develops. The controls of these various aspects of vascular differentiation are the subjects of Chapters 5 and 6; here only the correlation of this differentiation with organ development need be established.

There is also a close correlation between the presence of an organ and the formation of non-vascular tissues of the axis, such as the cortical parenchyma (Sachs, 1972a). These tissues, however, are formed at a very early stage of organ development and their development is not affected by the removal of relatively mature organs.

Another expression of the relations of an organ and the axis that connects it to the plant is the abscission or degeneration of axial tissues, depending on the organ and the plant (Fig. 2.4; Jacobs, 1962; Warren Wilson et al., 1986). Abscission or degeneration occurs, or at least is hastened, in an axis that connects with organs that have been removed, damaged or have ceased to develop (as in self-pruning). Conversely, both

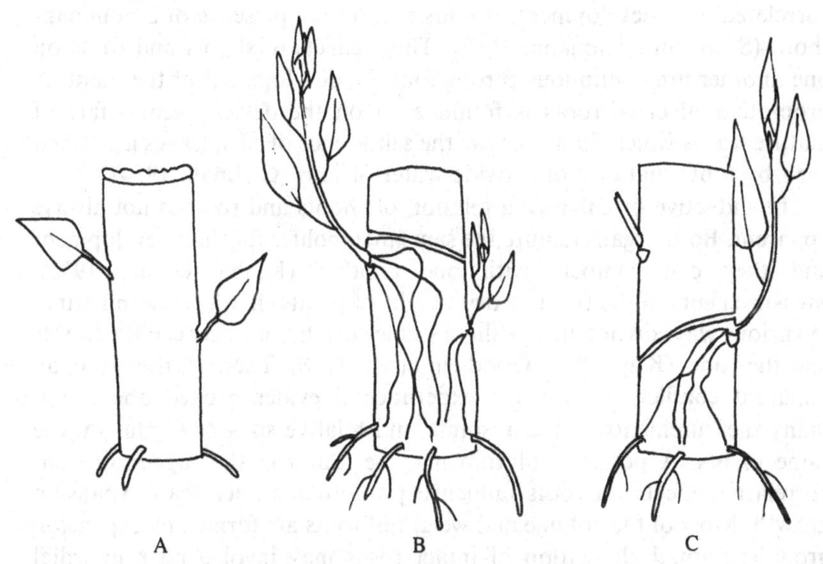

Figure 2.4. Summary of the inductive effects of a growing shoot on the rest of the plant. A. Control cutting with no growing buds. Such cuttings may form some roots and there may also be a development of a callus on the cut surface from which the original growing shoot was removed. B, C. Cuttings on which different buds develop into growing shoots. Bud growth is correlated with inhibition of the growth of other buds, differentiation of vascular tissues, increased initiation of roots and degeneration or even abscission of mature leaves and of the parts of the axis that do not connect the new growing shoot with the roots.

abscission and degeneration are inhibited when the axial tissues connect with a developing organ.

(d) Developing shoots enhance root development and vice versa

This relation is complementary to the inhibition of similar organs, the 'apical dominance' mentioned above. An inductive effect of shoot tissues on the initiation of root apices is seen most clearly in cuttings, which have no roots (Fig. 2.4). It is also demonstrated by experiments on plants where all shoots have been removed (Fig. 2.3B; Keeble et al., 1930; Wightman and Thimann, 1980). It thus appears that the inductive influence of shoot tissues can control the formation of roots.

Roots, on the other hand, are necessary for shoot development even under experimental conditions in which water and essential ions are readily available. This is seen in the initiation of new shoot apices on cuttings of various plants. Even when both types of organs are missing, root initiation may occur first, and it can be essential for the initiation of shoots (Harris and Hart, 1964). Extensive, branched root systems are correlated with development of buds even in the presence of a dominant shoot (Sachs and Thimann, 1967). These effects of shoots and roots on one another are continuous throughout the development of the plant. A promotive effect of roots is found even on the developmental fate of mature leaves which do not grow: the senescence of such leaves is delayed even by roots that do not provide water or ions (Chibnall, 1954).

The inductive or enhancing relation of shoots and roots is not always apparent. Both organs require the same metabolites for their development and they can compete with one another (Keeble et al., 1930). Measurements of the fresh or dry weight of plants that have been treated in various ways do not always show stable correlations between the shoots and the roots (Kny, 1894; Goodwin et al., 1978). There is, therefore, an apparent conflict between the experimental evidence cited above and many measurements of the absolute and relative sizes of organs on the same plants. A possible solution may be found in the suggestion that correlative effects on roots influence primordial rather than expansion growth. Most of the volume and weight of roots are formed by expansion growth; normal elongation of intact roots may involve no primordial development (Chapter 10). Separate controls of primordial and expansion growth of roots could therefore account for the apparent conflict between measurements of weights and experiments on the formation of new organs. Furthermore, the correlation between shoots and roots may be maintained by episodic corrections, so that it varies with time (Drew, 1982), and this correlation is also influenced by environmental conditions (Watts et al., 1981). There are some additional exceptions that could not be related to these various reservations: e.g. shoots that grow and branch in the absence of roots and vice versa, especially in organ cultures (Torrey, 1955; Wang and Wareing, 1979). Regardless of the basis of these cases, they are at most exceptions to the valid generalization that shoots and roots promote the primordial development of one another.

QUANTITATIVE EXPRESSIONS OF CORRELATIVE SIGNALS

The approach adopted above was to seek generalizations on the basis of a broad survey of developmental phenomena. This rules out precise quantitative relations, ones that could be subject to rigorous measurements and could be described by mathematical equations. Yet some statements are possible on the basis of descriptions of repeated correlations between developmental events, even when the quantitative relations are necessarily not linear or simple.

One conclusion from broad observations is that *the various responses to correlative signals are quantitative* rather than all-or-none. This quantitative aspect of correlative responses is seen most clearly in the orientation of differentiation towards a dominant organ. The extreme form of orientation is vascular differentiation, where responses to correlative signals differ in the numbers of cells of various types and their size, most obvious in the diameter of the vessels (Aloni, 1988b). Oriented differentiation also takes less-extreme forms: cell elongation and divisions at right angles to orienting influences (Sachs, 1984a). Quantitative rather than all-or-none responses to spatial signals are also seen in the promotion or inhibition of the number, size and rates of development of various primordia. Even in the case of abscission a quantitative aspect is apparent in the time that elapses before an organ is actually separated from the plant. Many physiological studies stress an all-or-none approach, but this is a result of a careful choice of experimental systems which is dictated by good reasons of experimental design, not an expression of the common reality of plant development.

A second generalization concerning quantitative correlations is that *the faster an apex develops, the larger are its effects on the rest of the plant*. The common expression of this is again the correlation of the vascular supply of an organ with its rate of growth. It is also true that apices which do not grow do not inhibit other apices. Lateral buds that have started growing, on the other hand, do inhibit the development of the main shoot (Libbert, 1955). Finally, rapidly growing leaves, not the ones that perform maximal photosynthesis, are the most inducive of both bud inhibition and root initiation (Snow, 1929).

An exact comparison of the various correlative effects of a primordial center requires arbitrary definitions. Yet in a general way the *Various correlative effects of an organ on the rest of the plant are quantitatively correlated.* A center of development that has a large influence on vascular differentiation also has large correlative effects on the inhibition and promotion of other organs. Where a correlative response of the plant, such as vascular differentiation or the formation of roots, is absent, it can be more readily explained on the basis of tissue competence than by the absence of a necessary signal. Thus in many plants mature tissues cannot form roots even when all the necessary signals from the leaves must be present.

Can any primordial development occur without the limiting signals of the rest of the plant? The removal of a growing shoot usually leads to cessation of the development events that were associated with its presence, such as cambial activity and the initiation of new roots. However, there are many exceptions where primordial development appears to start without an obvious inductive source. Some if not all of these exceptions could be due to problems of defining which tissues are sources of correlative signals. An assumption that *correlative signals originate in all tissues*, apices and other regions of rapid development being their major but not their unique sources, would lead to the conclusion that no primordial development occurs in the absence of inductive spatial signals.

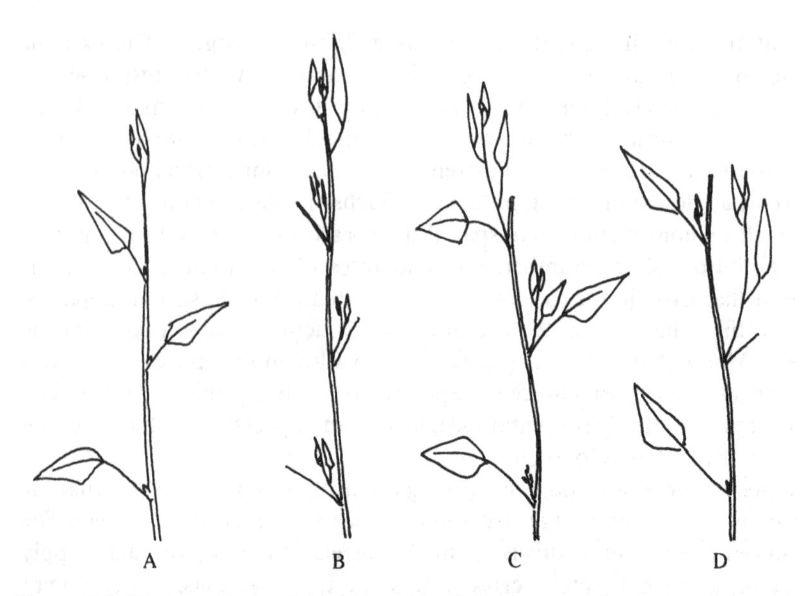

Figure 2.5. The correlative inhibition of bud development by mature leaves. A. Intact control in which the growth of the lateral buds is not readily seen. B. Limited bud growth occurs, even in the presence of the main shoot apex, where mature leaves have been removed. C. The removal of the main shoot apex results in pronounced growth of more than one lateral bud. D. As in C, but one mature leaf was removed. The bud in the axil of this leaf has a clear advantage over other buds and is most likely to replace the original, dominant shoot apex.

This attractive picture is supported by the effects of removing mature organs such as expanded leaves (Fig. 2.5; Dostál, 1909, 1967). But it is at best a working hypothesis which requires further research.

The evidence that correlative signals elicit quantitative responses does not indicate that they are the only control of development. In fact, growth, expressed by changes of size, dry weight or fresh weight, is readily influenced by environmental factors and is not consistently correlated with events in the rest of the plant. This growth is, however, largely due to increases of water, cellulose and lignin while the facts marshalled above concern the division and differentiation of cells and the formation and continued development of new organs. The conclusion reached above, that the correlative interactions between shoots and roots are most evident concerning the initiation of new organs, may be indicative of a broader generalization. It is *primordial development*, and not other aspects of growth, that is under direct control by interactions with the rest of the plant. This subject is taken up in later chapters.

PRIMORDIA ARE LIMITED BY THE DIVERSION OF CORRELATIVE SIGNALS

The generalizations above suggest relations that could be a first step towards understanding how various correlative controls of development are coupled or coordinated. The following statements are most important. (a) Signals received from the rest of the plant limit the initiation of shoots and roots. (b) New organs dominate the plant and orient tissue differentiation towards themselves; this domination presumably orients the transport of critical signals and all other resources towards these apices (Sachs, 1984a). Thus it can be suggested that the more active a new organ is, the more limiting signals it receives; these signals in turn induce further development (Sachs, 1972b, 1975b). A general working hypothesis would be: *The development of apices is limited by an exchange of signals with the rest of the plant; this exchange is the source of a positive feedback: the more rapidly an apex develops the more rapidly it is likely to continue developing.*

The dependence on correlative relations is often evident throughout the development of an organ: it can even be found in delays of senescence of determinant organs, such as leaves. There are, however, important exceptions in which the future of an organ is largely set in the early stages of its development. One such exception was mentioned above: correlative relations determine only the primordial growth of roots (Chapter 10). Other examples involve the differentiation, rather than the growth, of the lateral organs that are influenced by correlative relations only at the time they are formed (Wareing, 1978; Meins, 1986).

An interdependence of signals through their effect on developmental processes is not meant to imply a direct biochemical relation. It is, of course, possible that one signal could directly induce or inhibit the synthesis of another, but living cells are complex and there is no *a priori* reason to assume an absence of intermediate processes. The biochemical level of description, furthermore, is not the one sought here (Chapter 1). For the present purposes it is sufficient to treat the cells as 'black boxes' and to lump together various ways in which signal action could be complementary and interdependent.

Consequences of the hypothesis

A control of primordial development by its relations with the rest of the plant has varied implications which will be the background for discussions in following chapters. Yet some possibilities should be mentioned at this stage.

(a) *Development stability could be due to correlative interactions*
The relation of an apex with the plant could be expected to be relatively stable and not to respond to transient conditions. This stability could be due to three different aspects of the correlative interactions between apical development and the rest of the plant: (1) The orientation of tissues towards a dominant center would continue to direct signals and resources for some time after the induction of orientation ceases. (2) A lag can be expected between the arrival of substances in an apex and the formation of inductive signals. Thus the signals a tissue forms should be influenced by its past state. (3) Tissues of a rapidly developing apex are the most competent to respond to the signals of the rest of the plant. They would therefore consume or otherwise change the signals needed for similar differentiation elsewhere.

The correlative relations between an apex and the rest of the plant could thus be a basis for development normally continuing where it has started. Such stability is suggested by old results, showing that the more inhibited an organ becomes, the more likely it is to be further inhibited (Snow, 1931). It is also expressed by the common observation that adventitious initiation or differentiation are 'last resorts': in most plants they occur only when organized apices are absent. Adventitious initiation requires an establishment of new centers of interactions with the plant, without previous tissue orientation and signal formation, and such initiation could not be expected to compete for signals and resources with apices that were organized earlier.

(b) *The importance of tissue orientation*
The induced orientation of both differentiation and transport is a major characteristic of hypotheses that apical development is controlled by interactions with the rest of the plant. This orientation appears to be defined by the axis of the movement of the signals originating in apices. This induction of orientation is not hypothetical; it is rather a description of the differentiation and function of the vascular system (Sachs, 1981a; Chapters 5, 6). Oriented transport is known to occur, however, in mature vascular tissues. A dominant role of the suggested feedback would require oriented transport to be significant, over short distances, even in meristematic tissues (Sachs, 1986). This requirement is reasonable, but it has not been experimentally tested.

(c) *Few signals could suffice for many correlative relations*
It is often assumed that different correlative effects of an organ on the development of the plant must be the result of the activity of different signals (Went, 1938). However, the various developmental responses outlined above could be due to a small number of master signals (Fig. 2.6; Sachs, 1988c). The signals of an apex could orient differentiation in

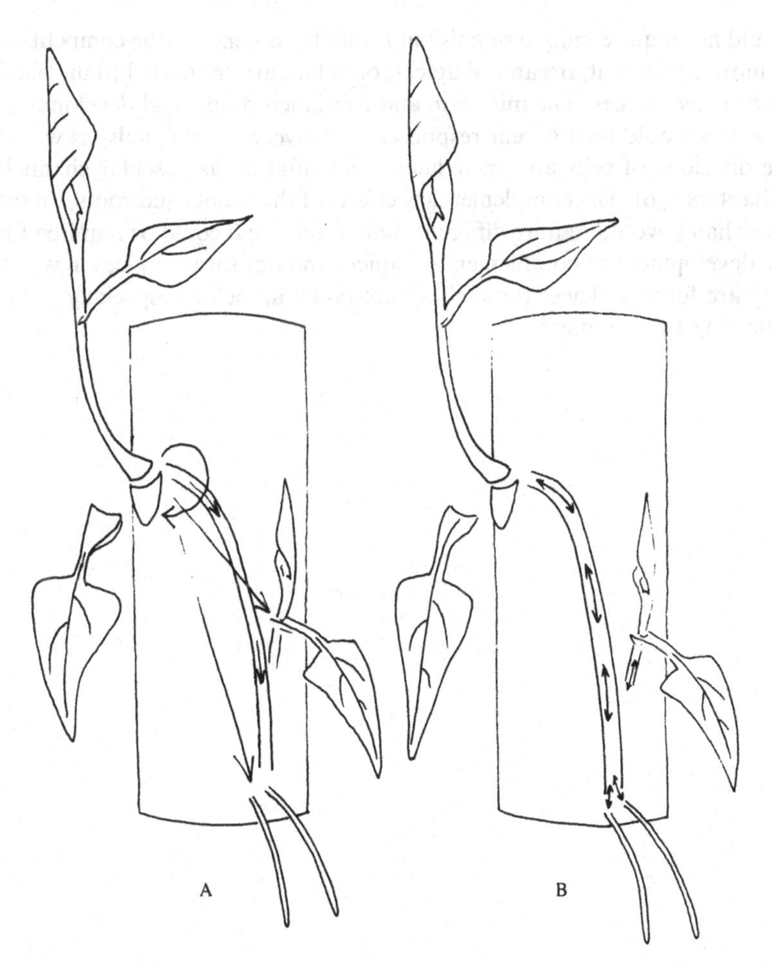

Figure 2.6. Representation of two possibilities concerning the varied effects of a growing bud on the rest of the plan. A. The growing bud is the source of different, specific signals inducing vascular differentiation, abscission of leaves, inhibition of growth of other buds and initiation of roots. B. The growing bud is the source of one signal which has varied effects: it induces vascular differentiation along the course of its polar flow and induces the initiation of roots where it accumulates. The influence of this signal on bud inhibition and abscission is indirect – it could be due to the diversion of other essential signals from the rest of the plant, and especially the roots, along the differentiating vascular tissues and towards the growing bud. The available evidence supports this second possibility.

accordance with the axis of their movement through the plant (Chapters 5, 6) and induce complementary initiation where they accumulate. There would be no need for special inhibitory effects, specific to different types of development – these could be due to the diversion of limiting signals away from weaker primordia (Overbeek, 1938). Adventitious initiation would not require unique signals but rather the absence of the competition of more competent, organized apices, ones that are connected to the plant by oriented tissues. The initiation and continued primordial development of apices would be different responses to the very same signals, as would the divisions of cells and their later differentiation as vascular channels (Chapters 5, 6). The complementary effects of the shoots and roots, on the other hand, would require different signals, ones that could be required for the development of complementary apices and not for the apices in which they are formed. These possibilities are taken up below, especially in the following two chapters.

3
Hormones as correlative agents

A source of auxin replaces the shoot apex in inhibiting the growth of lateral buds (Fig. 3.1). This was discovered soon after the identification of auxin as a hormone that controls growth in oat coleoptiles (Thimann and Skoog, 1933, 1934; Thimann, 1977). At the same time it was also shown

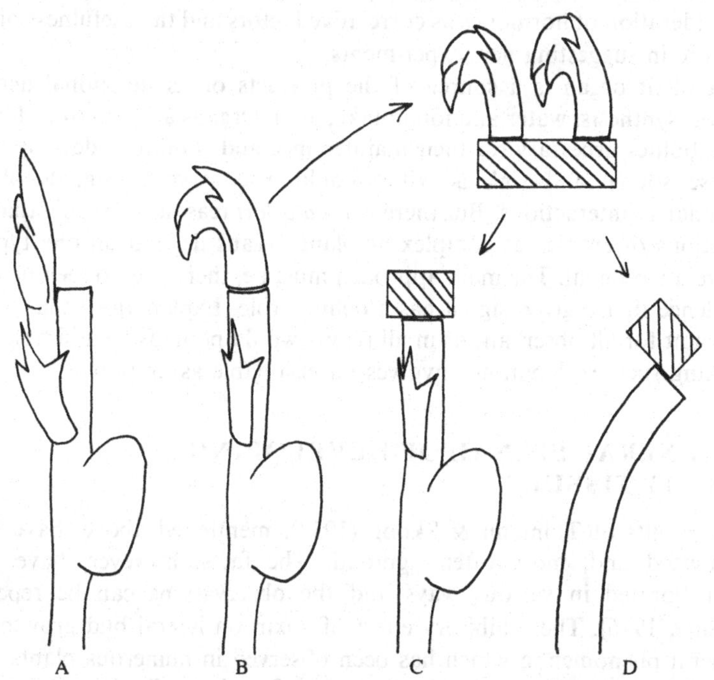

Figure 3.1. Thimann and Skoog's (1933) experiment concerning the signal by which growing shoot apices inhibit the growth of lateral buds. A. Removing the main apex from a *Vicia* seedling results in bud growth. B. Apices were cut and placed on an agar block. C. This agar partially replaced the intact apex in inhibiting lateral bud growth on decapitated plants. D. The very same agar had specific auxin effects in the *Avena* curvature bioassay. The experiment indicates that apices inhibit bud growth, that growing shoot apices are a source of auxin and that auxin at least partially replaces the inhibitory effect of apices.

that substances with auxin activity were present in the species used for the experiments (*Vicia faba*). Quantities of auxin comparable to those found in the plant were sufficient to cause some bud inhibition. On this basis it was concluded that *auxin is a signal of growing shoot apices* that inhibits lateral bud growth.

The significance of these results is that they identify a correlative signal. They raise the question whether the effects of auxin on bud growth are common to many plants and whether auxin and other known plant hormones have additional correlative roles. This chapter attempts to answer these questions and to formulate a general picture of the correlative relations between organs in terms of hormonal signals. The stress is on the role of hormones in organized development, not on the receptors and mechanisms of hormone actions on plant cells (Guern, 1987). The approach adopted here thus neglects many valid concepts and facts concerning plant hormones (Barlow, 1987). The justification for this neglect is the simplicity and clarity of the picture that emerges from the consideration of hormones as correlative factors and the usefulness of this picture in suggesting new experiments.

A plant organ is a source of the products of its functional activity (photosynthesis, water and ion uptake, etc.). Organs are also sinks for the metabolites necessary for their maintenance and continued development. These source and sink activities could also serve as signals of the correlative interactions. But there is no *a priori* reason to assume that the relations of entities as complex as plant organs depend on one type of correlative signal. The main approach must be, therefore, to seek positive evidence that a given signal has a defined role. Explanations that would account for all observations in all plants would perhaps be desirable, but seeking such explanations involves unreasonable assumptions.

A GENERAL SIGNAL OF DEVELOPING SHOOT TISSUES

The results of Thimann & Skoog (1933), mentioned above, have been contested and, more often, ignored. The facts, however, have been corroborated in various ways and the observations can be repeated (White, 1976). The inhibitory effect of auxin on lateral bud growth is a general phenomenon which has been observed in numerous plants. Not surprisingly, inhibition by auxin is not found in all plants under all possible conditions, though the exceptions appear to be rare (Jacobs et al., 1959; Thimann et al., 1971). Inhibition of bud growth by a dominant shoot is also more pronounced than inhibition by applied auxin (Thimann and Skoog, 1934; Wickson and Thimann, 1958). This difference could be due to the involvement of additional signals, mentioned below. The effect of exogenous auxin could also be limited by its being applied to the entire

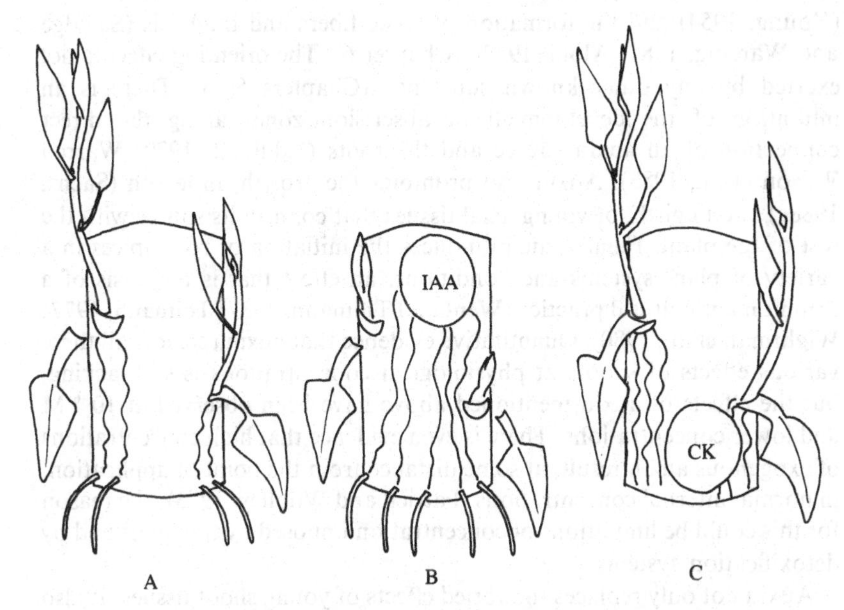

Figure 3.2. Developmental effects of auxin and cytokinin applied locally to cuttings. A. Untreated cutting. The growth of lateral buds, the initiation of roots and the differentiation of new vascular tissues connecting the new organs form a new, integrated plant. B. Locally applied auxin (IAA) induces the differentiation of vascular tissues. These vascular tissues follow the polarity of the plant and connect the point of auxin application with the roots. This same auxin also inhibits bud growth, induces the abscission of leaves and induces the initiation of new roots. These varied effects imitate the presence of the growing bud in the untreated cutting. C. The local application of a cytokinin (CK) inhibits the initiation of roots and enhances the development of buds. The same cytokinin also delays the senescence and abscission of the mature leaves. Cytokinins enhance but do not orient or localize the differentiation of the new vascular tissues. These varied effects of cytokinins imitate the presence of growing roots.

wounded surface rather than to the precise channels in which it is transported in the intact plant (Thimann and Skoog, 1934).

Auxins have effects other than bud inhibition and these are significant to any consideration of their role as correlative agents. The remarkable generalization is that auxin replaces all the correlative effects of a shoot apex (Fig. 3.2B; Sachs, 1975b, 1986, 1988c; Matthyse and Scott, 1984). These effects are specific to auxin, but they may be increased or modified by various additives (Hejnowicz and Tomaszewski, 1969; Lachaud, 1983). Thus, auxin prevents plagiogeotropic branches from turning upwards (Snow, 1945) and, when applied together with gibberellins, induces stolon growth in the lower buds of potato plants (Woolley and Wareing, 1972). Local exogenous auxin also induces and *orients* all aspects of vascular differentiation, with the possible exception of the early procambium

(Young, 1954) and the formation of some fibers and tracheids (Savidge and Wareing, 1981; Aloni, 1987b, Chapter 6). The orienting effect is not exerted by any other known substance (Chapters 5, 6). There is an inhibition of the development of abscission zones along the direct connection of an auxin source and the roots (Addicott, 1970; Warren Wilson et al., 1988). Auxin also promotes the growth, in length (Sachs, 1988c) and in girth, of young axial tissues that connect its source with the rest of the plant. Finally, auxin induces the initiation of root apices in a variety of plant systems and conditions, an effect that is the basis of a common agricultural practice (Went and Thimann, 1937; Thimann, 1977; Wightman et al., 1980). Quantitative evidence that auxin replaces all these various effects of shoots at physiological concentrations is still lacking, but the effects of auxin mentioned above have been observed at 10^{-6} M and lower concentrations. There is even evidence that high concentrations of exogenous auxin result, at some distance from the point of application, in normal internal concentrations (Patrick and Woolley, 1973); the reason for this could be limitations of concentration imposed by transport and by detoxification systems.

Auxin not only replaces the varied effects of young shoot tissues: it also has no other repeatable effects at the organ or whole plant levels. This variety and specificity of the responses to auxin could not be due to chance. It can be concluded that *auxins are signals by which growing shoot tissues influence the development of the rest of the plant* (Sachs, 1975b, 1986, 1988c). This is a logical conclusion, not an hypothesis. It follows from the facts that auxin is formed by shoots, is known to be transported away from the shoots, and that it specifically replaces the effects of shoots on the rest of the plant. The conclusion does not mean that auxin is the only correlative signal of shoots: a unique signal is neither likely nor required by the evidence. Finally, the conclusion is limited to the role of auxin as a signal leaving the young shoot tissues, stating nothing about its mechanism of action.

As suggested in Chapter 2, another important conclusion follows if the observed effects of auxin reflect physiological processes: *the various correlative effects of shoots do not require separate signals*. The choice between different developmental processes would then depend on the competence, and thus on the previous developmental history, of the responding tissues; it would also depend on whether the signals flow through the cells or are concentrated in them (Chapter 6). The quantity of auxin could still indicate the size and rate of development of the shoot above and the flow of auxin could specify direction, important in the control of growth of young tissues and of vascular differentiation (Chapter 5), even though auxin is a general, unspecific signal and is not formed in any one type of cell.

There have been many objections to this conclusion about the correlative role of auxin. A thorough discussion of these objections would require an entire book, but they stem from one or more of the following expectations which are not justified on logical grounds. (a) There should be only one signal for each correlative relation. (b) A signal can only be known if its mechanism of action is understood. (c) A signal must elicit one specific cellular process. (d) The effects of auxin, if it is a correlative signal, must be related in a simple way to the measurable concentrations of known auxins. This simple relation should not be expected, especially not in view of the extraction procedures that must be used (Davies, 1987b; Chapter 4). (e) Finally, signal changes in a tissue must be unique and not immediately followed by all the changes that necessarily occur during growth and differentiation.

It is interesting that auxin was discovered on the basis of a working hypothesis that there must be a substance that specifically controls the process of cell elongation. It was soon discovered that auxin has many other effects on plant development, and this led to the suggestion that it might not be a 'real' hormone (Went, 1938). The conclusion reached here is that the useful working hypothesis that led to the discovery of auxin was wrong. There is nothing 'typical' about the effect of auxin on extension growth. The essential role of auxins in the plant is characterized by their origin in developing shoot tissues, not by any developmental process they elicit (Sachs, 1975b, 1986, 1988c).

HORMONAL SIGNALS OF ROOTS

The correlative influence of the roots on the rest of the plant is in various ways complementary to that of the shoots (previous chapter). It is hardly surprising, therefore, that the influence of removed roots cannot be replaced by local sources of auxin. Correlative effects of roots, however, may still be related to the flow of auxins that generally originates in the shoots (Chapters 5, 6). It is therefore possible that roots influence the plant by acting as sinks for auxin (Sachs, 1968). Such sink activity of roots would be expected to inhibit additional root initiation. This initiation is known to occur where auxin accumulates, in the absence of sufficient 'sinks'. Even if roots do act as sinks for auxin, however, they could still be active sources of hormonal or other signals.

Cytokinins were discovered as factors necessary for cell division in culture (Skoog and Miller, 1957) and the enhancement of cell division has been regarded as their 'typical' function. As with auxins, however, it was soon discovered that cytokinins have varied effects on plant tissues (Figs. 3.2c, 3.3; Matthyse and Scott, 1984). Thus cytokinins inhibit the initiation of root apices in many plant tissues and under varied conditions (Skoog and Miller, 1957; Wightman et al., 1980). Cytokinins have the opposite influence on shoot apices: they promote their adventitious initiation and,

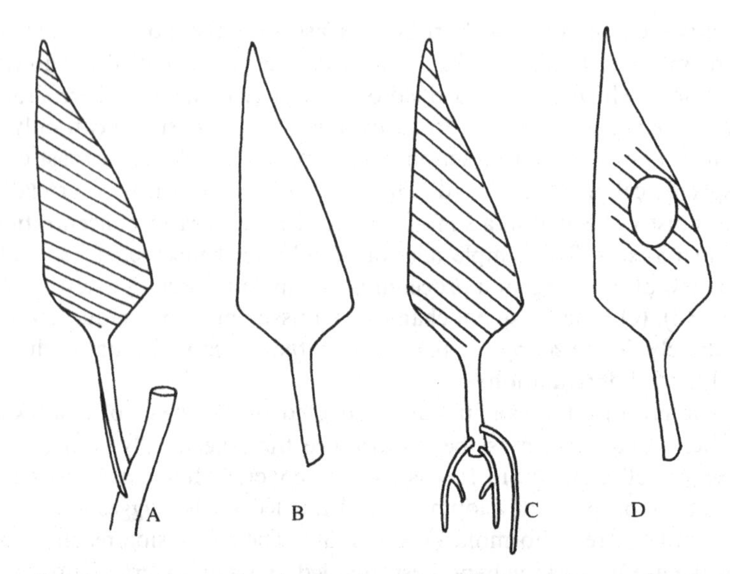

Figure 3.3. An experiment showing that locally applied cytokinins replace roots in maintaining leaf activity. A. Leaves on a decapitated plant, which has no growing buds and thus no young leaves, remain active for a long time, even for a number of years. B. A similar leaf cut from the plant turns yellow and undergoes senescence, generally within a week. This is true even when water and essential ions are provided. C. A similar detached leaf remains active for an indefinite period if roots develop on its cut petiole. D. Local applications of a cytokinin at the site shown imitate the roots in delaying leaf senescence.

when shoot apices are already present, promote their growth even in the presence of an inhibitory shoot or an exogenous source of auxin (Skoog and Miller, 1957; Wickson and Thimann, 1958; Sachs and Thimann, 1967; Sossountzov et al., 1988). Cytokinins also counteract inductive effects of dominant shoots on lateral shoot apices: They cause lateral branches of potato plants to turn upwards and develop large leaves (Woolley and Wareing, 1972) and they prevent the senescence of inhibited pea shoots (Sachs, 1966). Local applications of cytokinins to leaves delay senescence; this delay is most apparent in detached leaves, with no connections with roots, where senescence is most rapid (Fig. 3.3; Richmond and Lang, 1957).

These effects of cytokinins might appear as a bewildering array of unrelated processes. Yet as in the case of auxin the picture becomes simple when they are considered as a signal of an organ and not a molecule that controls one master process: cytokinins replace various effects of roots on the rest of the plant. It is also known that roots are a major source of cytokinins in plants (Kende, 1965; Goodwin et al., 1978; Feldman, 1979). Since cytokinins are formed by roots and replace their effects, it may be suggested that *cytokinins are a major correlative signal of developing roots* (Sachs, 1975b, 1986, 1988c). Quantitative evidence that the amounts of cytokinins that are produced by roots influence the rest of the plant is, on the whole, still lacking. Cytokinins do, however, influence plants at low, 'hormonal' concentrations (10^{-6}M) and in very small quantities (Sachs and Thimann, 1967). The suggested role of cytokinins in the plant does not depend on their being formed exclusively in the roots. Nor does this conclusion depend on cytokinins replacing all the effects of the roots – at least one other correlative effect, the action of the roots as sinks for auxin, was mentioned above.

HYPOTHESIS CONCERNING THE HORMONAL CORRELATIONS BETWEEN PLANT ORGANS

It was concluded in the previous chapter that there are developmental correlations between the shoot and root systems of a plant and that these correlations are expressed not only by the development of the various apices but also by the differentiation of the vascular tissues that connect them. It was concluded above, in the present chapter, that auxin and cytokinins are major signals of the shoots and roots. It may be suggested, therefore, that *there are hormonal interactions involving auxins and cytokinins that control various relations between the parts of a plant* (Fig. 3.4; Sachs, 1975b). The existence of these interactions would depend on the following general principles.

(a) Hormone synthesis, or at least its release, is quantitatively related to the rate of apical development. The relation need not be linear and it need not involve all aspects of development.
(b) The development of an apex is limited by the supply of the hormone it must receive from the rest of the plant. This, too, need not be true in all conditions nor for all aspects of development. It is likely that development is associated with the consumption or change of the limiting hormone that arrives from the rest of the plant.
(c) The amount of hormones and other limiting substances which an apex receives from the plant depends not only on its present but also on its past developmental activity. This would guarantee a stability of apical development.

To the extent that these relations hold, auxin formation would depend, indirectly, on the supply of cytokinins and vice versa. A growing apex would influence the rest of the plant both by being a source of its characteristic hormone and by acting as a sink for the signals of the complementary apices. This dual control could be important in preventing malfunction. It could also prevent the expression of somatic mutations that would cause the formation of excess signals by organs, in no

Figure 3.4. Diagramatic representation of a hormonal interaction between the shoot and root systems of a seedling. This interaction is much more apparent in primordial development than in growth and increase of dry weight. The root is first to grow and it is the source of cytokinins, and presumably other signals, that enhance the apical growth of the shoot. The newly formed leaves are the source of auxin which induces the primordial root development, especially the formation of lateral branches. This interchange of signals continues, and it maintains an overall balance between the primordial development of the shoot and the root. Because of transport time and lag periods the development of the shoots and roots can be out of phase, though not necessarily to the extent shown in the figure.

relation to their development and environment. The supply of limiting hormones to an apex would be a function of its enhancement of the development of complementary organs and inhibition of competing, similar ones. Another aspect of the same relations is that an apex would polarize the axial tissues, leading to the continued differentiation of transporting tissues. This could be expected to lead to a diversion of supplies towards the dominant apices.

The hormonal hypothesis offers a simple picture of inter-organ relations and it has a number of interesting theoretical features. If true, it would confirm the suggestion made in the previous chapter, that very few signals account for the various effects of an organ on the rest of the plant. Even the diversion of limiting hormones towards the dominant apices could depend on the same correlative effects, since these effects include the induction of oriented vascular differentiation. This would mean that the very same signals could be responsible both for the correlative inhibition of similar development and for the stability of the continued development of the apex that is the source of the inductive signals. The origin of these signals in different parts of the plant and limitations of their supply would be major controls of organization. The importance of this principle at a tissue level is suggested by the development of tumors when auxin and cytokinins are formed within the same tissues (following chapter).

Limitations of the evidence

A major aspect of the hypothesis concerning the role of auxins and cytokinins is the localized and separate synthesis of organ-specific hormones. There is good evidence that auxin synthesis occurs primarily in the shoots and cytokinin synthesis in the roots (Avery, 1935; Scott and Briggs, 1960; Sachs, 1975b; Goodwin et al., 1978; Matthyse and Scott, 1984). It also appears that the expanding tissues, the ones with the largest correlative influence, are the major sites of synthesis. Mature tissues, however, may also be sources of both hormones and quantitative generalizations about the precise localization of synthesis appear impossible (Chapter 2). The relative role of different tissues probably varies, depending on the genotype, environmental conditions and developmental stage. It is also possible that some auxin synthesis occurs in the roots, at least when they are cut from the plant. This would be one way of accounting for root branching in culture, in the absence of added auxin (Torrey, 1955). Cytokinins may be formed in shoots, though the best evidence for this is again for cut plants (Wang and Wareing, 1979). These and similar cases may be exceptions or indications of an absence of strict localization of hormone synthesis. In neither case would the hypothesis developed above be disproved as a general principle. It is clear, however, that much more quantitative evidence is called for.

There are also hardly any data concerning the assumed relations between hormones and the rates of development (Goodwin et al., 1978). One reason for this lack of evidence is that it has not been sought. The expected dependence of auxin synthesis on cytokinin supply and vice versa, however, would be indirect: it would operate through the controls of rates of development. The question of the occurrence of a hormonal control is therefore independent of the unknown molecular mechanisms of hormone action. Furthermore, observations of qualitative relations between rates of development and inductive effects, such as those expressed by vascular differentiation, do suggest that measurements would show the expected quantitative relations. There are also a few indications that additions of one hormone to cultured tissues can promote the formation of another (following chapter).

There is no doubt that substances are diverted towards growing apices and even towards mature tissues to which hormones have been applied (Gunning and Barkley, 1963; Patrick, 1976). It is less clear how much of this is due to these meristematic tissues acting as sinks (Brenner, 1987; Morris and Arthur, 1987) for available metabolites and how much to the controls of transport (Patrick and Wareing, 1976) and the induction of oriented differentiation (Gersani et al., 1980a,b). An orienting influence on transport could also act as a mechanism of correlative inhibition: it could prevent limiting substances from reaching the inhibited apices (Overbeek, 1938; Gregory and Veale, 1957; Sachs, 1970, 1981a; Morris and Winfield, 1972; Raju et al., 1978). One objection to the idea of such inhibition by diversion has been that its only clear expression is vascular differentiation, and this is a slow process. Oriented transport, however, could take place even at the early stages of cell polarization and the presence of vascular tissues need not be an invariable indication of transport capacity (Kirschner and Sachs, 1978; Sachs, 1984a). Various other objections to this 'diversion hypothesis' are based on the assumption that there could only be one mechanism of apical dominance (Sachs, 1981a).

ROLES FOR ADDITIONAL DEVELOPMENTAL SIGNALS

Auxins and cytokinins may be major signals of shoots and roots, but they are hardly likely to be responsible for all the correlative relations of something as complex as plant organs. And in fact there is plenty of evidence that developmental processes depend on a balance of a few or even many factors, some of them being hormones. Controls are said to be exerted by complex networks, requiring a quantitative systems approach. The conclusions reached above are in sharp contrast; they assign clear roles to two known hormones. The difference, of course, is not in the facts but rather in the questions asked. It was assumed above that in a living plant any given process could be influenced by many factors and the discussion is therefore limited to clear comparisons of events in the presence and in the absence of one source of controlling signals, a plant organ. This class of questions is preferable *because* it yields a simple picture.

There is evidence concerning substances, such as polyamines (Smith, 1985) and oligosaccharins (Albersheim and Darvill, 1985) which influence development but are not known to have correlative roles. Electrical signals are known in plants (Pickard, 1973) and could serve as correlative signals, but the evidence for a wide role is not conclusive. A lot is known about the role of other plant hormones, which change according to environmental conditions and modify almost all aspects of development (Matthyse and Scott, 1984; Davies, 1987a,b), but the available evidence concerning the role of factors other than auxins and cytokinins in the organ correlations considered here is at best suggestive. There are, of course, many reports of effects of known hormones on organ development. These are important, especially concerning the responses of plants to their environment, but they do not include evidence that the hormone replaces the influence of an organ and that this organ is known to be a source of the growth factor. Such evidence is essential for any serious suggestion concerning the nature of a correlative signal (Jacobs, 1959). There is also little evidence that the reported effects are common to different plants and to varied conditions. Thus the evidence that substances are 'involved' in apical dominance shows that their absence, or presence in excessive amounts, is detrimental to normal development. This might mean that the substances are important controls, but not that they necessarily have a role of correlative signals.

There are, however, many indications that require further study. For example, ethylene or its precursors may be signals of waterlogged roots (Jackson and Cambell, 1975). Abscisic acid (ABA) may well be a signal of stressed roots (Blackman and Davies, 1985). Indications that could be especially valuable for further research concern gibberellins, especially in relation to their effects on apical differentiation (Chapter 12), their replacement of root effects on the plant (Carmi and Heuer, 1981), and their specific effects on fiber differentiation (Hess and Sachs, 1972). Other indications come from organ culture. Various organic substances, whose role is not known, are often required in the culture media (Murashige, 1974). Thus, thiamine and myo-inositol are generally necessary, though they are not needed for the growth of intact plants nor are they known to influence wounded plants. It is possible that such substances have correlative roles associated with their being produced in mature tissues and transported towards meristematic centers (Bonner, 1942).

There are various results concerning organ correlations that cannot be explained on the basis of hormonal controls. For example, damage of cotyledons of *Bidens* seedlings leads to changes in their axillary buds within less than a minute (Desbiez et al., 1984). This is an extreme example of responses that are much too rapid to be accounted for by known hormonal signals. Other rapid responses include a measurable increase of remaining bud size following shoot decapitation (McIntyre and Damson, 1988). This might be a response to the change in water status of the plant which may or may not be related to later bud growth. A final example is results showing that correlative inhibition can cross through non-polar, zigzag tissue bridges in cut plants (Snow, 1938). These and other facts all require further study; they presumably show that correlative relations between organs depend on more than one simple control system. This is an important conclusion, but it does not disprove the positive evidence summarized above concerning the role of auxins and cytokinins as a major, though not necessarily unique, control of various relations between plant organs.

Metabolites as correlative signals

Metabolites and essential ions are transported in the plant and their availability often limits growth. It has been shown that such substances move preferentially towards dominant organs (Loeb, 1924). This, however, is hardly more than would be predicted from macroscopic observations of development. There is also clear evidence that the addition of growth-limiting factors, such as essential ions, reduces apical dominance (Thimann et al., 1971; McIntyre, 1977). On the basis of these facts it has been concluded that the movement of substances towards the region in which they are used for growth serves as a signal of correlative inhibition (McIntyre, 1977). This is possible and deserves further study; an example of a result that is specially suggestive is the promotion of bud growth by the local application of K^+ (Wakhloo, 1970). Yet a general role of metabolite distribution as a primary correlative signal is doubtful for the following reasons:

(a) The movement of essential substances would account for correlative inhibition, but not for the induction of vascular differentiation and other correlative effects that, as considered above, are associated with the same controls.
(b) Essential substances would be poor signals since they would not be specific to certain types of organs and certain developmental processes. Their effects would be most pronounced in relation to mature or expanding organs rather than primordial development.
(c) The available results can be readily understood in other ways. Conditions that promote growth would also be expected to influence

the formation of hormones. The availability of limiting metabolites could thus act on organ correlations via an effect on growth rate and hormone formation.

(d) The responses to environmental changes are often too rapid to be accounted for by the availability of metabolites, water or ions. Thus, decreases in cytokinins have been found to precede actual changes in mineral supply (Kuiper, 1988) and increases in ABA precede water shortage in drying soils (Blackman and Davies, 1985).

All this does not rule out the possibility that metabolite availabilities serve as secondary controls, ones that become dominant in specialized conditions. There is no logical necessity to consider signals and metabolites as alternatives rather than components of a complex control system.

4

Callus and tumor development

Cells close to a wounded surface often divide. These divisions result in a tissue known as *callus*: a mass of relatively large cells without obvious shape or organization (Figs. 4.1, 4.2; Küster, 1925). The growth of callus generally ceases after a short time, when the wound is covered and the various parts of the plant and their vascular connections have regenerated. Occasionally callus growth does continue and large masses of tissue are formed. Such callus, whose growth is unlimited, is a *tumor* comparable to animal cancer (Figs. 4.1, 4.2; DeRopp, 1951; Braun, 1978). Unlimited callus growth also occurs in *tissue culture*: when parts of the plant are isolated on a medium that provides all the substances necessary for growth (Gautheret, 1959; Murashige, 1974).

The previous two chapters dealt with the correlative relations of the development of organized structures. It is now necessary to ask about the correlative relations of unorganized development, the development of callus and tumors. Is this development related to, and influenced by, the hormonal signals involved in organ relations? This question is significant because any positive answer, partial though it may be, would indicate that long-distance interactions between organs could also be important for the cellular organization of tissues.

DIFFERENTIATION AND PARTIAL ORGANIZATION OF CALLUS AND TUMOR TISSUES

It is often assumed that callus represents a mass of 'undifferentiated cells'. However, simple microscopy shows that cellular heterogeneity is very common (Fig. 4.2; Gautheret, 1957, 1959). Most callus cells are parenchymatous: they are relatively large, thin-walled and have a large vacuole. The parenchymatous cells divide, but there may also be groups of cells with the characteristics of apical meristems. Tracheary elements are the specialized cells most commonly reported, but this may be due to the ease of observing their thick lignified walls. Sieve elements are noticed only upon careful study (Aloni, 1980), so it is not clear whether their occurrence is widespread. Fibers are rare in callus and various specialized

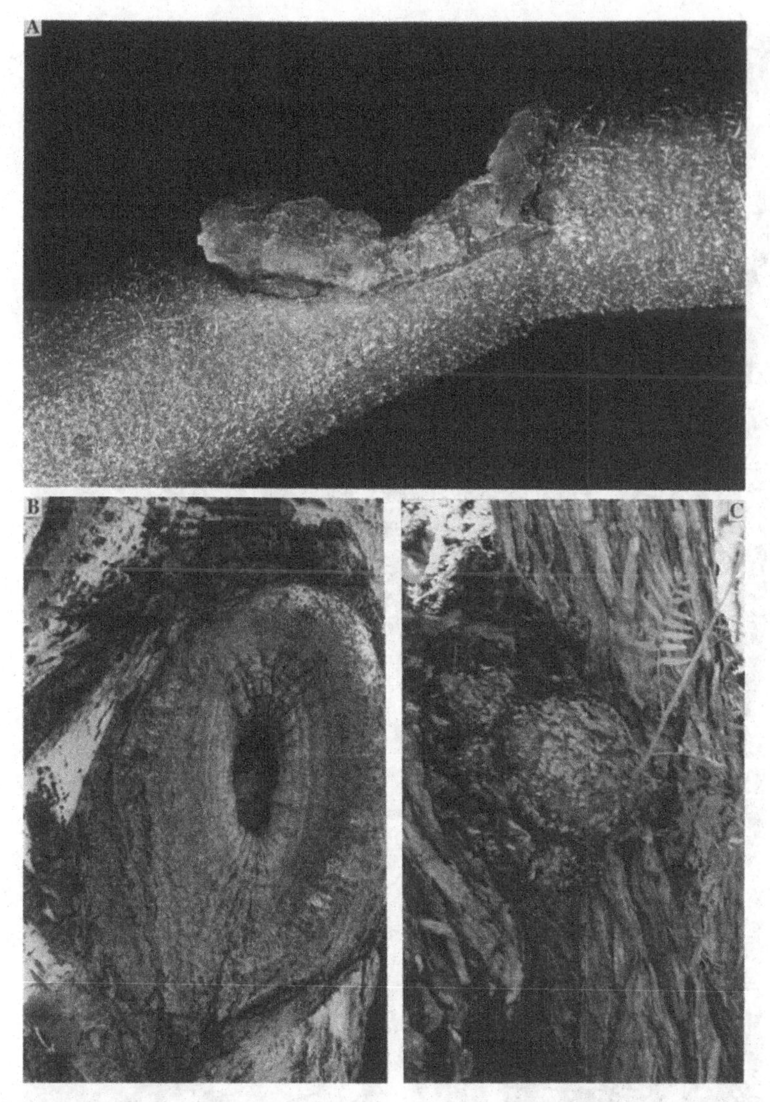

Figure 4.1. Callus and tumor development on various plants. A. Callus development on a cut surface of a bean hypocotyl about 10 days after it was wounded. The hypocotyl was cut from the shoot and the roots but left in contact with the cotyledons. It was kept in a moist atmosphere in unsterile conditions. B. Callus surrounding a cut where a large branch was removed from a poplar tree. There is considerable organization within such callus. C. A large tumor on a *Schinus molle* tree. The cause of these tumors is unknown, but they often form where the trees have been wounded. Their development is indefinite, though organized shoots may appear, apparently from within the tumor tissue.

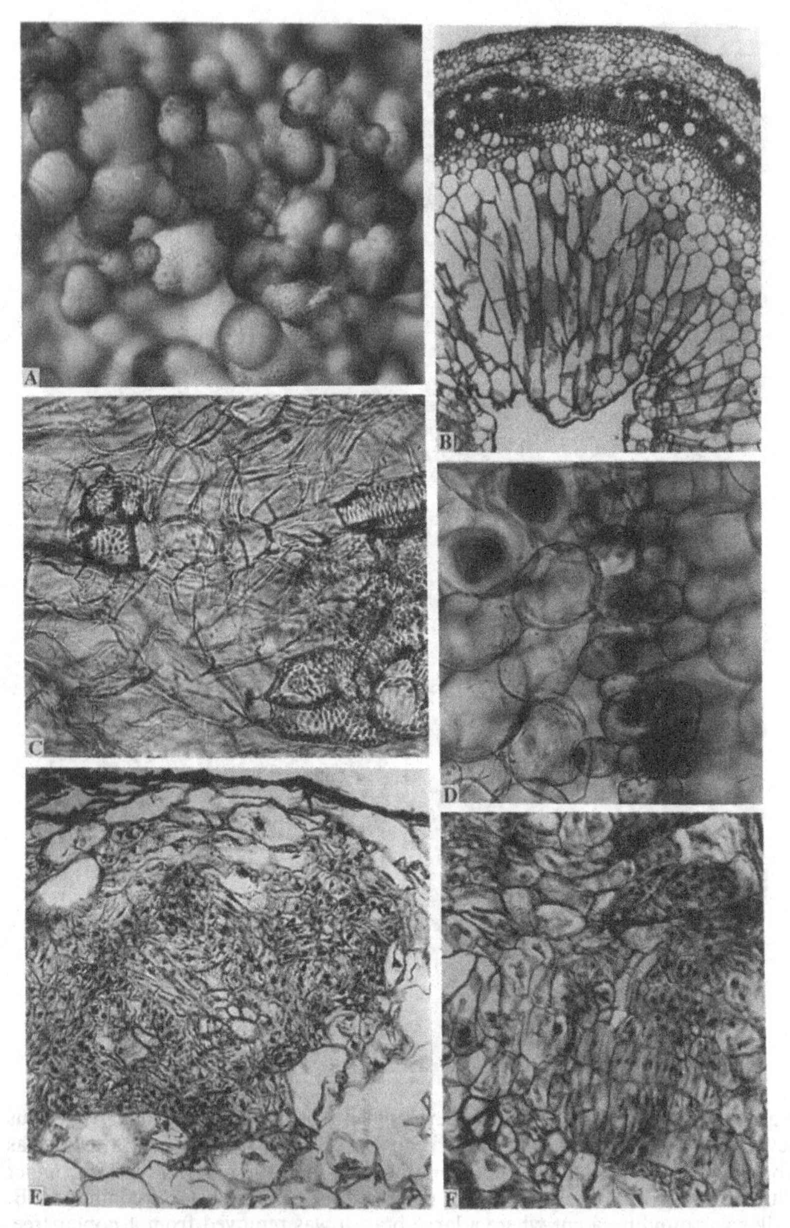

Figure 4.2. The cellular structure of callus of various sources. The pictures demonstrate that callus is generally not composed of uniform cells and it can have varied degrees of organization. A. Surface view of callus formed on a cut stem of a pea seedling which was kept in a saturated atmosphere. The cells were heterogenous; part of this heterogeneity was due to the origin of the callus from different tissues. X100. B. Cross-section through a bean hypocotyl about 10 days after it had been severely wounded. The rapid new growth was relatively

cells formed by meristemoids (Chapter 7) are apparently absent. Though the cellular heterogeneity found in callus is not invariable, it is definitely found in many tumors, both on plants and in culture (Noël, 1946).

Stable or determined differentiation of callus cells is expressed also in the dependence of their traits on the tissue from which the callus originated (Heslop-Harrison, 1967; Murashige, 1974; Wareing, 1978; Meins, 1983, 1986; Wareing and Al-Chalabi, 1985). This dependence of callus traits is found not only for callus originating in different organs but also for different degrees of the juvenility to maturity gradient along shoots (Chapter 12). The differentiation of the callus is recognized in its form, rates of development, regenerative formation of new apices, hormonal requirements for growth (Meins and Lutz, 1979), and even in the antigenic properties of cell surfaces, which are maintained during repeated sub-cultures (Raff et al., 1979; Meins, 1986).

The absence of normal organization, rather than the lack of cell differentiation, is the distinguishing characteristic of callus (Murashige, 1974). This need not mean that callus is necessarily uniform and devoid of all organization. Varied partial organization is seen even in the external form of callus: both on the plant and in culture there are often domes that must be centers of greater growth (Fig. 4.1). The size and location of these domes vary, but for any given callus they are not quite random. It is possible to find all intermediates between such growth centers and organized leaves and roots. These intermediate, partially organized structures are known as *teratomata* (Braun, 1953; Turgeon, 1982). Teratomata occur in healthy tissues, though they are best known as special types of crown galls, considered below. Masses of partially organized roots are also common, in both tumorous and normal callus tissues. Perhaps the least organized callus is suspensions of cells growing in liquid cultures, but even in these the individual cells are often elongated and thus have one preferred axis.

All degrees of partial organization can be found in the distribution of the various cell types. Callus consisting only of elongated cells, all oriented

---

organized: it had a clear axis, at right angles to the surface of the wound. X 20. C. Thick section of callus formed at the base of a cut bean hypocotyl kept in a saturated atmosphere. Some of the cells of the callus differentiated as tracheary elements, cells that are dead when they are mature. The organization of these tracheary elements was not along any one dominant axis and could not be followed in sections. X 100. D. Early stage in the formation of callus on a cut stem of *Impatiens sultanii*. The line of dense, small cells formed from the cambium. These cells were quite different from the ones that developed from the adjoining parenchyma. X 110. E, F. Sections through a crown gall on *Helianthus annuus*. Hypocotyls of seedlings were wounded and infected with cultures of *Agrobacterium tumefaciens*. The tumor is composed of differentiated cells of various types. Cells of any given type bear repeated relations to other, similar cells: the tissue is partly organized. E, X 200; F, X 120.

in the same way, seems to be typical of the early stages of growth following wounds. In some plants and tissues, this oriented growth may continue for some time (Fig. 4.2B). Sooner or later, if there is continued growth, it involves the appearance of many developmental centers. When there are meristematic cells they are arranged in groups. The presence of many centers of development results in tortuous axes of differentiation. The most common expression of these varied axes may be the unusual, irregular shapes of many callus cells (Fig. 4.2). The most obvious indicators of organization are tracheary elements. In sections these elements often appear scattered, with no relation to one another (Gautheret, 1959). But in a 3-dimensional view, in cleared material, tracheary elements are commonly arranged in strands (Sachs, 1975b). The axes of the strands follow tortuous, complex curves and they often branch. Even the nodules of vascular tissue common in cultured callus could be twisted, branched strands with no clear beginning or end (Sachs, 1981a). But since nodules cannot be cleared, reconstructions of 3-dimensional structure from serial sections are required – and these are not yet available.

It may be concluded that the controls of callus development are not merely a matter of local divisions and growth rates. Callus is not undifferentiated, not uniform and not completely lacking in organization. It is at least likely that the internal structure of callus is related to the conditions that cause callus growth.

CALLUS AS A RESPONSE TO DISRUPTED CORRELATIVE RELATIONS

Callus development occurs when plants are wounded. This could mean that callus-inducing substances ('wound hormones') leak out where cells are damaged. Such substances might also be formed after wounding, in chemical reactions in the crushed cellular materials. But wounds could also be expected to have another effect: they could disrupt the distribution of correlative signals by interrupting their passage between the parts of the plants. (Chapters 2 and 3). These interruptions could result in inductive substances becoming concentrated in relatively mature tissues near the wounds. These two possibilities, wound hormones and disrupted organization, are not mutually exclusive, since both could contribute to the development of callus.

Great stress was once placed on the hypothesis of 'wound hormones'. Evidence for their existence was found in the reduction of callus growth when wounded tissues were washed thoroughly. It is not clear, however, that such washing necessarily removes specific substances; it is just as likely that washing inhibits all growth by removing ions or other essential factors. One bioassay for the activity of 'wound hormones' was the

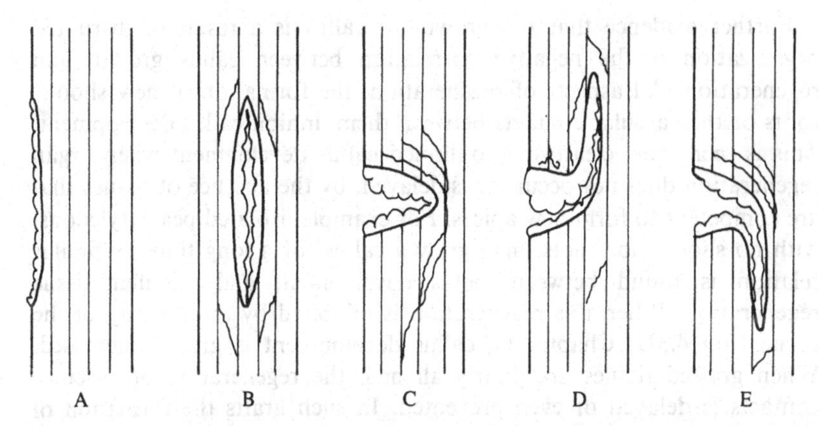

Figure 4.3. The relation of callus growth to the depth and orientation of wounds. Diagramatic representation of experiments performed on hypocotyls of bean seedlings. A. Shallow wounds, which do not disrupt the vascular system, caused only limited callus development. B, C. The influence of wounds depended on their orientation. In B the wound was parallel to the plant axis and caused only minor and temporary disruptions of the vascular contacts. In C the wound was at right angles to the axis and it cut many vascular strands. Though these wounds may have killed the very same cells, callus development was generally greater in C, where the disruption of correlative relations was more pronounced. D, E. Combined transverse and longitudinal wounds. When these led to a polar dead-end (D) callus development was most pronounced. Tissue flaps of the opposite orientation (E) did not present a 'trap' for polar transport and they did not have the same effect on callus development.

growth of callus on damaged bean pods. Using this bioassay a callus-enhancing substance, called 'Traumatin', was isolated (Bonner and English, 1938). This substance has not been found to have widespread, repeatable effects on plant development and its role in wounded tissues is not clear (Zimmerman and Coudron, 1979). Thus, available evidence does not support the hypothesis of specific 'wound hormones' though it is quite likely that non-specific, growth-promoting substances are released when something as complex as a living cell is cut or crushed.

In contrast, there is direct evidence that the disruption of the intercellular relations that maintain normal organization is an important factor causing callus development. Cuts made in various orientations in bean hypocotyls, for example, may damage the very same cells and yet they result in very different rates of callus development (Sachs, 1981a; Fig. 4.3). Transverse cuts are the ones that sever the major interactions along the plant axis (Chapter 5) and it is these cuts that result in the most pronounced callus growth, at least in bean hypocotyls (Sachs, unpublished). Differences in the effect of cuts with varied orientations could depend on correlative interactions, but not on the damage or on the nature of the cells that first start growing.

Further evidence that the growth of callus is a result of disrupted organization is the negative correlation between callus growth and regeneration. All aspects of regeneration, the formation of new shoots, roots or the vascular contacts between them, inhibit callus development. This is seen most clearly by continued callus development when organ regeneration does not occur, or is delayed, by the absence of tissues that are competent to form new apices. For example, isolated pea cotyledons, with no shoots nor roots, may grow a callus for a long time. A similar relation is found between callus development and vascular tissue regeneration. When this regeneration is inhibited by the polarity of the tissue (Fig 4.3D; Chapter 5), callus development is greatly increased. When grafted tissues are poorly aligned, the regeneration of vascular contacts is delayed or even prevented. In such grafts the formation of callus is specially pronounced (Sass, 1932).

Finally, the disruption of correlative relations can be imitated by the addition of known hormones. Callus development is enhanced by the application of auxins, cytokinins or ethylene. A most effective enhancer of callus development is the auxin 2,4-dichlorophenoxyacetic acid (2,4-D). This is a synthetic substance, so that excess amounts would not be detoxified as readily as natural auxins, and transport away from the point of application would be polar but not rapid (McCready, 1963). Furthermore, unlimited callus development is possible without repeated wounds – and thus without any 'wound hormones' – over a period of years. This occurs, as considered below, in tissue cultures and in tumors when the necessary correlative factors, auxins and cytokinins, are continuously present, either because they are added to culture media or because they are continuously produced by the tissue itself.

AUXIN AND CYTOKININ REQUIREMENTS OF CULTURED CALLUS

The source of the substances required for callus development must be the plant or, in culture, the medium. It follows that the components of media permitting continued growth could be relevant to the nature of callus tissues and to their relations with the plant. The study of the chemical interactions between the parts of a plant was, in fact, one of the purposes of originators of tissue culture methods. The success of culture work and its varied, unexpected uses (Vasil, 1981) have caused this aspect to be virtually forgotten. The significance of media composition should be considered now that chemically defined media are available and the growth of practically all plant tissues is possible (Murashige, 1974).

Culture media (Reinert, 1982) always include a carbon and energy source, generally sucrose. They also include a balanced mixture of salts which supply the obligatory component atoms of living matter. All these are expected and do not add to our knowledge concerning the relations of callus with the rest of the plant. Of greater interest are a small number of organic molecules necessary for the culture of most tissues but not required by intact plants. These substances include representatives, either natural or synthetic, of only two groups of known plant hormones: auxins and cytokinins (Skoog and Miller, 1957; Murashige, 1974). In addition, a small and varied number of other molecules are necessary, the most common being thiamine and myo-inositol (Murashige, 1974). It is not known whether these additional substances have correlative roles in the intact plant (Chapter 3). It can be concluded that the number of substances that serve as promotive signals and are received by a growing tissue from the rest of the plant is quite small. Chief among these substances are representatives of the two hormone groups found to be important correlative signals in the relations between organs (Chapter 3).

Habituation: a differentiation of callus expressed by its hormonal requirements

As mentioned above, the requirements for continued growth in culture differ depending on the source of callus (Meins, 1986). But there are also stable changes that occur during the continued growth of callus. Gautheret (1947) found that repeated transfer of callus led to stable variants that grew well on a medium devoid of auxin. This 'Habituation' is due to the formation of auxin by the callus itself and has appeared in various laboratories (Wyndaele et al., 1988), but it is rare or at least unpredictable. A related habituation, more amenable to experimental work, is the changed requirement for cytokinins from the media (Meins, 1982, 1983; Meins and Wenzler, 1986). This habituation for cytokinins is not an all-or-none trait: different levels of cytokinin requirements can be readily found. The frequency with which cytokinin habituation appears and the possibility of its reversal show that it is an epigenetic or differentiation event; mutations, selected to become dominant in culture, could not be expected to behave in the same way and they can be ruled out by the instability of habituation during regeneration and sexual reproduction (Meins and Binns, 1979, 1982; Meins, 1982, 1983). These facts demonstrate a quantitative differentiation of plant cells concerning their hormonal requirements, a differentiation which occurs even in cultured callus.

Habituation resembles other cases of differentiation in depending on the origin of the callus (Meins and Wenzler, 1986; Jackson and Lyndon, 1988) and, at least in some cases, in the possibility of its induction by specific hormonal treatments (Meins and Lutz, 1980; Christou, 1988). A stable habituation in cultured leaf tissues of tobacco is also controlled by a Mendelian gene (Meins et al., 1983; Meins and Wenzler, 1986).

Habituation for cytokinins may reflect partial organization of root apices within the callus (Kerbauy et al., 1988). There is also some evidence that habituation may occur in organized apices, not only in cultured callus (Leshem and Sachs, 1986; Gersani et al., 1986), leading to teratomatous structures. The indications, in this and other chapters, that hormonal relations have a major role in determining normal development suggest that habituation may represent an important control event that occurs regularly during normal development; but this possibility requires much more study.

A HORMONAL BASIS OF TUMOR DEVELOPMENT

Tumors were characterised above as callus that is able to continue growing, even after regeneration has repaired any damage caused by the wound that caused the original callus growth. It follows that tumors are able to withstand inhibitory effects of growing organs even after the plant has had time to regenerate (Sachs, 1975b). This does not mean, however, that tumors are invariably insensitive to all such correlative effects: growth may continue in the presence of dominant shoot apices and still be increased when these apices are removed (Skok, 1968). There appear to be quantitative differences between tumors in this regard, but these differences have apparently not been documented.

In culture, tumor tissues are special in their ability to grow even in the absence of added auxins and cytokinins (Braun, 1956, 1958; Gautheret, 1959). This appears to be due to their ability to form these substances in unregulated amounts. The autonomous formation of growth substances is the one universal trait that has been found in various tumor tissues of independent origins (Braun, 1978), and it is significant that this is the only cellular characteristic by which tumor tissues of varied origin have been found to differ from normal cells. Furthermore, the Ti plasmid of crown gall bacteria is a direct cause of tumor development (Hooykaas et al., 1982) – and although this plasmid includes varied genetic information, the part of this plasmid essential for the tumorous conversion of cells is concerned with the synthesis of auxins and cytokinins. A possible reservation is that measurements do not always yield clear correlations between concentrations of hormones and unorganized growth (Weiler and Spanier, 1981; Pengelly and Meins, 1983). It is possible, however, that such measurements may not indicate the activity of substances present in critical compartments. It has also been known for a long time that a source of auxin replaces many of the effects of a tumor on the plant (DeRopp, 1947): it induces a large vascular strand, whose polarity is that of a contact with a growing root apex (Fig. 4.4c), inhibits shoot growth and induces root initiation (Sachs, 1981a). Additional important evidence for the role of hormones in tumor development is that the application of relatively large quantities of auxin and cytokinin to normal plant tissues causes them to develop as though they were tumors, both in culture and on a wounded plant (Sachs, unpublished).

How could the unregulated synthesis of two correlative factors, auxins and cytokinins, be responsible for the development of tumors and their ability to continue growing even on regenerated plants? The conclusions of the previous chapter do suggest an hypothesis (Sachs, 1975b). Auxin orients and induces the differentiation of vascular contacts that transport metabolites to the tumor. The internal supply of cytokinins make the tumor independent of the normal limitations of the development of auxin sources on the plant. Thus *the unregulated synthesis of only two known hormones is a cause of unorganized growth in plants*. Neither here nor in other chapters are the molecular basis of tumor induction and the possibilities of its reversion considered – these important topics relate to the control of differentiation and not directly to the controls of patterning or developmental organization (Chapter 1).

Varied synthesis of auxins and cytokinins is also the basis of differences of tumor structure. Crown galls may be in all possible variations between leafy teratomata, relatively unorganized callus, and masses of deformed roots. When a given plant and bacterial strain are used the tumor morphology still varies: as with normal callus, considered below, this morphology depends on the location on the plant relative to the various organs (Braun, 1953; Gresshoff et al., 1979; Turgeon, 1982). As might be expected, teratomata with many deformed leaves are most likely on the side of a cut that is in contact with roots and not with shoots. Variations of tumor form are also dependent on the strain of the inducing bacteria, and the genetic information that is the direct cause of crown gall development, the Ti plasmid (Hooykaas et al., 1982). The variations of tumor morphology are the expected ones in relation to hormone synthesis: relatively high cytokinins and low auxin levels are associated with unorganized shoots and the inverse hormonal relations cause a callus that includes many root apices (Amasino and Miller, 1982).

CORRELATIVE EFFECTS OF CALLUS TISSUES

According to the discussion above, callus growth is a response to the disruption of the correlative relations between the various parts of the plant. Tumors are a special type of callus – they are able to take part in correlative relations and to dominate the plant on which they are growing. This raises the possibility that even normal callus could be an active, though weak, participant in the correlative relations of the plant. A corrolary of this suggestion is that there could be varied forms of callus, whose correlative roles are different and even complementary. Evidence concerning these possibilities is limited, but the subject is directly related

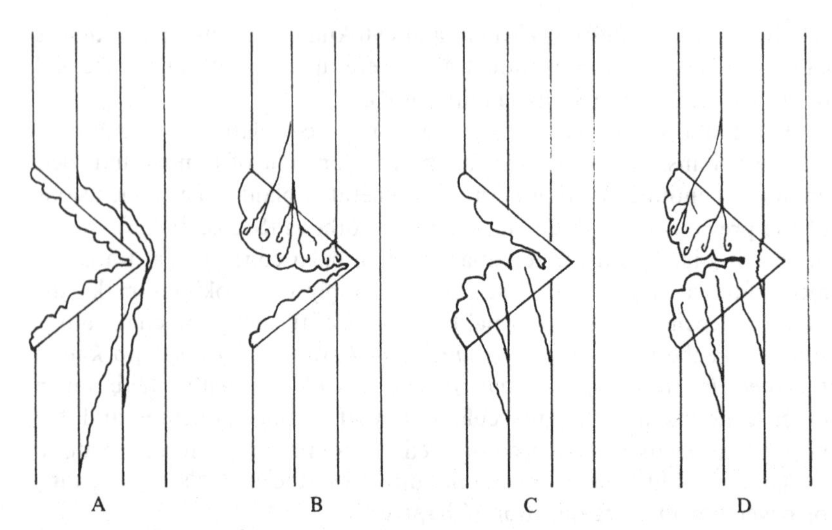

Figure 4.4. The vascular contacts of callus tissues on wounded bean plants. The wounds were the same in all cases and differences in callus development were due to environmental conditions, to the age and thus the competence of the wounded tissues and to the presence or absence of various organs on the plant. A. Vascular differentiation was absent from callus whose development was limited. This was true when the callus was limited in size or when the callus had large cells and large air spaces, and was thus limited in cell number and cytoplasmic content. B. Callus on a basal side of the plant, morphologically in the location of roots, was associated with pronounced vascular differentiation. The strands had direct contacts only in the direction of the leaves, were branched and finally ended in vessels and sieve tubes in the form of whirlpools. C. Callus on the apical side of a wound, replacing the shoot, had characteristic vascular contacts and vascular structure: the strands follow tissue polarity towards the roots; their contacts with the leaves, and thus with the sources of photosynthetic products, were always indirect. The strands themselves were fairly straight and they did not include vascular whirlpools. D. Both types of callus, replacing the shoots and the roots, may develop on opposite sides of one wound.

to tissue organization and this justifies a short review of the available facts.

The very growth of normal callus tissues must mean that callus influences the plant, since it is able to direct essential metabolites towards itself. A common, though not invariable, expression of the same or a related influence is the differentiation of oriented vascular contacts between the callus and the rest of the plant (Fig. 4.4; Kirschner et al., 1971). Callus development can occur with no vascular differentiation but, though no clear, quantitative statement is possible at present, the absence of vascular tissues seems to occur only in callus that is limited in quantity or has numerous intercellular spaces and large cells with very little cytoplasm. But when vascular contacts do form they suggest further information about the effects of callus on differentiation. As in the plant as a whole (Chapters 5, 6), the vascular differentiation associated with callus development connects with either the direction of the shoots or the direction of the roots (Fig. 4.4; Kirschner et al., 1971). This means that in relation to vascular differentiation the callus acts as a partial replacement of either the shoots or the roots (Chapter 5; Kirschner et al., 1971; Sachs, 1981a). The vascular contacts of crown gall tumors are always in the direction of the roots, and contacts of these tumors with leaves above them are necessarily indirect. This is true even when the tumor is at the base of the plant – the 'crown' of the root – and all the photosynthetic leaves are above. There are indications that non-tumorous callus at the base of cuttings, the one that replaces roots, resembles roots in delaying the senescence of leaves (Wheeler, 1971). The available data are limited because the topic has not been studied, but also because the effects of the callus, unless it is tumorous, are quite weak.

Callus organization is another expression of the relations of callus with the plant and of differences between two types of callus. The structure of callus formed on the one side of a transverse wound is generally different from the structure of the callus that develops on the opposite side from very same tissue (Fig. 4.4D; Simon, 1908; Küster, 1925; Priestley and Swingle, 1929; Bünning, 1953). Only callus that is connected to shoots, and therefore replaces roots, includes circular vessels and vascular nodules (Sachs and Cohen, 1982). This is the structure expected when there is excess auxin that has no polar outlet (Chapter 5). Other callus on the basal side of plant cuttings includes many deformed roots, again a known response to auxin accumulation (Mott and Cure, 1978). Callus on the shoot side of cut plants often has a different structure: it includes tissues oriented along parallel axes with no clear focal points (Sachs, 1981a). Judging from its effects on vascular differentiation, this apical callus is a source of auxin (Kirschner et al., 1971; Sachs, 1981a). Because of tissue polarity (Chapter 5), this auxin is readily drained towards the rest of the plant, so the absence of vascular nodules in the callus is readily accounted for.

Further evidence that callus can partially replace either roots or shoots comes from its interactions with intact, developing organs. Though the regeneration of wounded plants inhibits callus development (see above), this development may still be enhanced when only shoots or roots are present on the plants. Callus growing where root regeneration would be expected is promoted by the presence of shoot apices. The development of this callus is reduced when young leaves are removed, even though these leaves could hardly be a source of major metabolites essential for growth. Callus that replaces missing shoot apices, on the other hand, grows more when roots are present than when they are removed. The enhancement of callus by specific organs cannot, as yet, be supported by clear measurements – the statements above were based only on observations of callus on various plants, especially bean seedlings (Sachs, unpublished).

The discussion here has treated callus as a relatively passive tissue, responding in expected ways to internal or external hormones. This may be an oversimplification: callus may well be a dynamic entity, taking part in feedback relations. Callus may not only form hormones but do so according to the conditions it is in, in response to hormonal and other signals. This important possibility has hardly been studied but there are various indications that the syntheses of auxin and cytokinins are dependent on one another (Ishikawa et al., 1988; McGaw et al., 1988; Palni et al., 1988; Wyndaele et al., 1988). Meins (1982) has found evidence that cytokinin supply induces cytokinin synthesis in cultured callus. An indication more in line with the hormonal hypotheses developed in the previous chapters is a loss of auxin requirement following cytokinin treatment (Einset, 1977), which could be due to an induction of auxin synthesis in the callus (Wardell and Skoog, 1969; Syono and Furuya, 1972). Thiamine synthesis may also depend on hormonal treatments (Dravnieks et al., 1969). Similar responses of callus could account for the results showing that tumors formed in response to 'defective' strains of crown gall bacteria develop as though they were forming missing hormones (Ishikawa et al., 1988).

CONCLUSIONS CONCERNING CALLUS DEVELOPMENT

The facts are varied and perhaps confusing, but they do suggest a general hypothesis. Callus is another expression of development that is dependent on hormonal interactions between the various parts of the plant. Callus development occurs when plants are damaged, leading to disrupted, unusual correlative relations. This disruption causes relatively mature, non-meristematic cells to resume growth and cell divisions. These cells form callus, with varied degrees of partial organization. In some plants and conditions the same cells also form new apical meristems, and when these meristems appear callus development is inhibited.

Callus tissues are not only a response to disrupted correlative relations; they can also partake in the correlative relations of the damaged plant, respond to inductive influences and form hormones of either the shoots or the roots. But the effects of callus are always weak: callus is thus unable to compete with intact organs once they are formed and their vascular contacts regenerate. Tumors are the exceptions: they are callus that has changed so that it is able not only to compete with organized tissues but actually to dominate the plant. All known conversions to the tumorous state require the unregulated formation of, or response to, auxins and cytokinins, the signals of shoots and roots. These hormones are formed within one tissue, independently of interactions with the plant. The unregulated synthesis of both auxins and cytokinins enables tumors to dominate the plant and, at the same time, to be insensitive to the limitations of essential hormones supplied by the rest of the plant.

The disruption of correlative relations is reflected not only by the location and development of callus but also by its internal, cellular organization. The partial organization of callus varies greatly, depending on different correlative relations with the plant and on the developmental history of the cells from which the callus had originated. Callus, whether tumorous or not, can be an unorganized mass of growth centers, with many conflicting axes of development and no clear dominance. Callus does have the capacity for oriented differentiation, even though the plurality of developmental centers results in orientations that are difficult to follow. This suggests the hypothesis that competition for limiting signals could be a determinant of tissue organization and not only of the relations between plant organs (Chapter 3). This is a rather far-fetched hypothesis, but it is indicated by the unorganized development of normal tissues in culture, where all required substances are supplied in excess. Some evidence for this extension of principles that apply to the relations between organs to the organization of tissues will be found in the following chapters.

It might be useful to underline these statements by pointing out what callus and tumors are not. There are no types of differentiation that are unique to callus. Nor is callus one uniform type of tissue: instead it is a class characterized by aberrant tissue organization. The relations of the cells of callus need not involve unusual signals. This is true even of a tumor, a type of callus that is abnormal and can be considered a disease.

5

The polarization of tissues

A. THE VARIED EXPRESSIONS OF POLARITY

Vöchting (1878; see Sinnott, 1960) pointed out the implications of the regeneration of new organs on branch cuttings (Fig. 5.1). It is a general rule that roots form close to the basal side – the region that was originally closest to the roots. Shoot localization is often less clear, but there is a tendency for shoots to be initiated, or more commonly to develop from

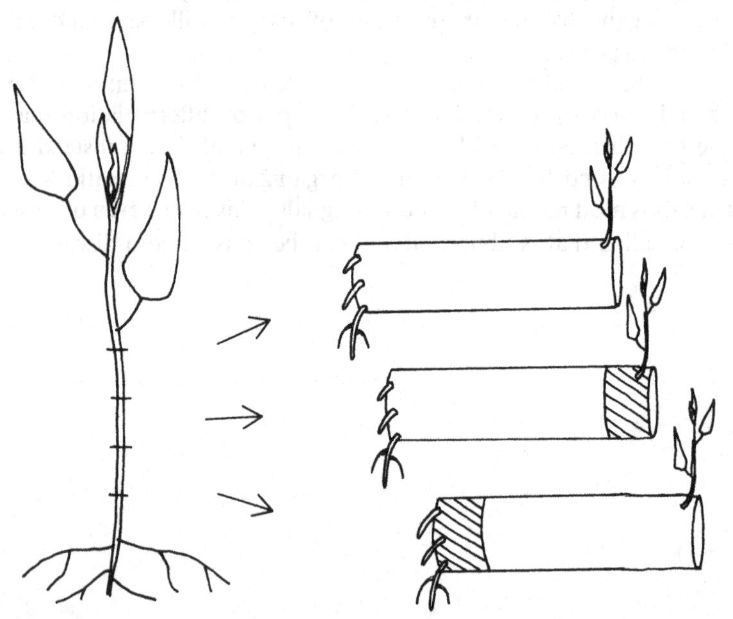

Figure 5.1. Evidence for a polarity of plant tissues: root regenerate from the base of a cutting and bud growth occurs primarily on its apical side. Cuttings taken along a plant axis all exhibit the same polar regeneration, even when the cuttings are kept horizontal and in uniform conditions. Development occurs in accordance with location on the cutting, not on the intact plant: the region marked by diagonal lines in the middle cutting forms buds; if the stem were cut differently, the very same tissue could be present at the lower side of a cutting and it would then form roots.

dormant buds, on the side opposite the roots, the side that was closest to the original shoots. There are exceptions to these generalizations, but Vöchting's results can be repeated readily using various plants. The localization of new organs is thus far from random and it expresses a directionality, or polarity, of a developmental process.

Vöchting's conclusion was that regenerative phenomena demonstrate a polarity of plant tissues themselves, not an expression of a response to an environmental factor. He further suggested that each cell is polarized – it is in some way analogous to a small magnet – and that polarity of tissues is only the sum of the polarities of its cells. A preferred axis and a preferred orientation are basic, essential expressions of patterned development, so the evidence for Vöchting's suggestions must be considered in view of present knowledge. The purpose here will be to classify various types of polarity and to consider the evidence for their role in plant tissues. The second part of this chapter deals with the induction and stability of tissue polarity.

TYPES OF POLAR DEVELOPMENT

A classification of possible polarities

The term polarity has been used in varied ways (Bloch, 1965; Sander and Nübler-Jung, 1981) so a classification of possible polarities is required. Four types of polarity which have the advantage of covering all possibilities are shown schematically in Figure 5.2. The four polarities are defined by answers to the following three alternatives. (1) A manifestation of polarity depends on factors either outside or within the developing tissue itself. (2) Polar development which depends on factors outside the tissue could be a function of either the environment or of the effects of boundary structures of the plant itself, such as roots and shoots. (3) Finally, if polarity depends on traits of the tissue, then it could be of two types (Sachs, 1984a): the cells may be unequal even in the intact plant, in which case some traits could be arranged along a gradient; or the cells could be initially equal but each cell could be polar – as suggested by Vöchting – transporting some critical signal in one preferred direction.

There is hardly need to consider how factors outside the tissue could localize the regeneration of roots and shoots, but the cellular gradients and polar transport require some explanation (Fig. 5.3; Sachs, 1984a). A gradient of the capacity to form roots could result in the first roots being initiated on one side of the cutting. These roots could prevent further root initiation and development by correlative inhibition (Chapter 2). An oriented or polar flow, on the other hand, could serve as a vectorial property along the entire tissue. When the plant is cut, signals transported along this polarity would accumulate on one side of the cutting. This accumulation could result in localized development. Thus, when polarity

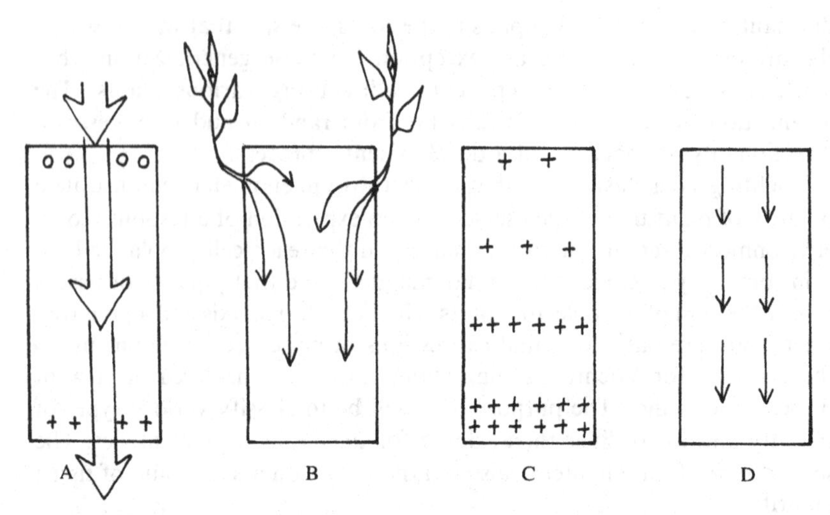

Figure 5.2. Four possible types of tissue polarity. A. The difference between the two sides of the cutting is imposed by an external, environmental factor – such as gravity or directional light represented by the arrow. B. Tissue orientation is imposed by growing organs, in this case developing buds. C. There is a stable gradient within the tissue. The gradient could be a difference in the concentration of a critical substance or of a state of the cells. D. Every cell is polar, polarity being expressed, for example, by an oriented transport of critical substance. A characteristic of this last polarity is that all the cells are originally equal.

is due to a gradient there are differences along the tissue even before the tissues are cut and these differences are accentuated by correlative inhibition. In the case of polar flow, the tissues could be uniform until the actual cutting causes the formation of a specialized region, where a signal accumulates and roots are initiated.

Evidence concerning different polarities

It is now necessary to turn to the evidence for actual roles of the various types of polarity in plant development. Since the various possibilities outlined above (Fig. 5.2) are not mutually exclusive, the question to be tackled cannot be which of these possibilities is correct. Instead, positive evidence must be sought for the actual occurrence of various types of polarity.

(a) The environment – gravity, light, humidity, etc. – could readily act as a cue for the localization and orientation of developmental events (Fig. 5.2A). There is no lack of evidence that such factors influence the formation and growth of root and shoot apices (Sinnott, 1960) and can be a cause of polar development. In general, such environmental effects on development are outside the scope of this book (Chapter 1).

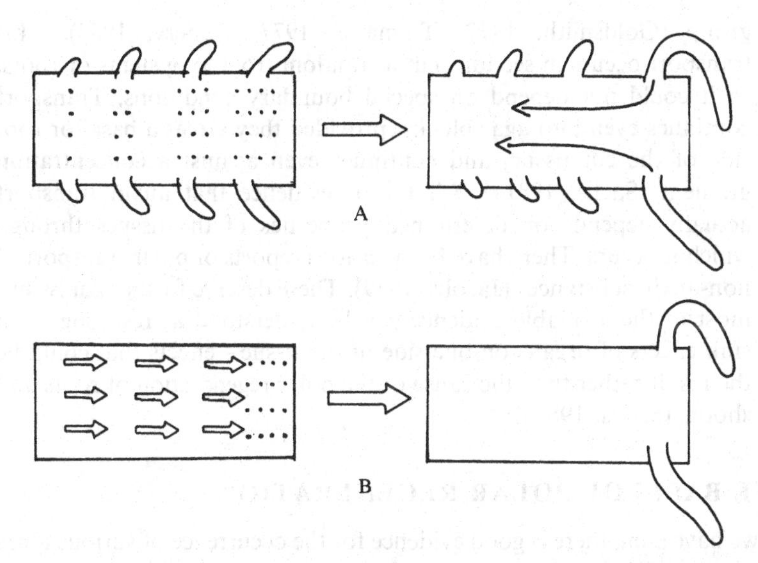

Figure 5.3. Possible mechanisms determining the polar regeneration of roots. A. Roots are initiated along the entire tissue. Because of an original gradient or pre-pattern within the tissues, root initiation and development occur faster the closer the tissue is to the basal side. After the roots are initiated the largest basal roots inhibit the continued development of all other roots, this being an expression of correlative inhibition. It is this last interaction which localizes the roots that are actually seen. B. Polar transport by all cells concentrates a critical substance at the basal side of the cut tissue. This localization induces the initiation of roots. In contrast to the possibility shown in (A), there is no pre-pattern of traits along the tissue and correlative inhibition need not play an essential role.

(b) The movement of metabolites, such as sucrose, through stems or roots is often along preferred direction and has therefore been considered to be polar. This directional movement could be the function of specialized regions, such as root and shoot apices, which act as sources and sinks for metabolites (Fig. 5.2B; Gersani et al., 1980a). Thus, directional transport need not reflect any innate polarity of plant tissues through which transport takes place – any more than a directional flow of water indicates directional traits of pipes (Sachs, 1984a).

(c) There is a variety of structural gradients in plants (Fig. 5.2C; Bloch, 1965). Gradients of developmental potentialities that could be most relevant to the topics considered here could be due to the gradual formation of the plant axis, as a product of apical growth. The tissues along the axis thus differ in their age, but also in traits that depend on gradual changes in the apices themselves (Chapter 12).

(d) Polar transport which is independent of the location of sources and sinks (Fig. 5.2D) has been demonstrated for substances of the auxin group (Goldsmith, 1977; Thimann, 1977; Rubery, 1987). This transport occurs in sections cut at random from long stems or roots, so it could not depend on special boundary conditions. Transport continues even into agar blocks, provided they are at a basal or root side of the cut tissue, and continues even against a concentration gradient (Sachs, 1981a). All this is evidence that auxin transport actually depends on determined properties of the tissues through which it occurs. There have been various reports of polar transport of non-auxin substances (Jacobs, 1979). These deserve further study, but most of the available evidence can be understood as resulting from sink effects of organs on one side of the tissue – effects that could be the result rather than the cause of the polar regeneration of roots and shoots (Sachs, 1984a).

THE BASIS OF POLAR REGENERATION

As we have seen, there is good evidence for the occurrence of various types of polarity (Fig. 5.2). Therefore, their relative roles in the regeneration of roots and shoots should be considered. Local environmental conditions can certainly determine the location of new roots and shoots. But this could not be the whole story of polar regeneration. For example, Vöchting's observations concerning the localization of new roots and shoots can be repeated on horizontal plant sections kept in a uniform environment (Fig. 5.1): they show that although environmental factors can be a cause of polar development, there must also be a polarity within the regenerating tissues themselves. It is interesting from an historical point of view that this conclusion, which appears obvious now, was not readily accepted and there were very long arguments concerning whether the polar localization of regenerative organs was a function of environmental conditions or of an internal tissue polarity (Vöchting, 1906). Apparently at the time it was assumed that only one clear answer could be true.

Some regenerative polarity must therefore depend on traits of the regenerating tissues: there are either gradients of these traits or there is a polar transport of critical signals (possibilities (c) and (d) in the previous section). Again these two possibilities are not mutually exclusive, so the evidence supporting the occurrence of each type of tissue polarity must be considered separately. There is evidence for tissue gradients which could influence organ formation. Thus, the gradient of tissue age does influence the rate at which new roots are formed, but the effects of this gradient are limited to the immediate vicinity of apices. Furthermore, in shoot cuttings roots are formed first at the basal cut, in the region that is most mature – i.e., in the location that is the opposite of that predicted by the effects of the gradient of tissue age. A second gradient is in the degree of tissue juvenility (Chapter 12). Since juvenility is associated with facilitated root formation, this gradient would predict that roots form at the base of cuttings, where they are in fact found. The gradient of root-forming capacity, however, is not steep: there is often only a small difference in the rate at which roots form at the base of cuttings taken along a long stem (Fig. 5.1; Sachs, 1984a). Furthermore, although quantitative studies are not available, there is no doubt that appreciable correlative inhibition is exerted only by organs that are growing rapidly and have a considerable mass (Chapter 2). Thus, small root primordia could not be expected to inhibit the early initiation of additional primordia higher up along a stem. It follows that the known gradient from juvenility to maturity could not account for the sharp polar localization of even the earliest stages of root initiation (Sachs, 1984a).

In contrast, there is concrete evidence for a role of polar transport in polar regeneration (Gautheret, 1944). Polar transport necessarily leads to an accumulation of auxin at one predetermined side of cut sections. Where polar transport cannot continue and auxin can be expected to accumulate, new roots are induced (Chapter 3). Applied auxin has the same root-inducing effect (Went and Thimann, 1937; Thimann, 1977). When the exogenous auxin concentration is high, so that it presumably cannot be carried away by polar transport, it readily causes the formation of new roots in unusual location, i.e., ones that are unexpected on the basis of tissue polarity.

It is not possible to state unequivocally that auxin and its specialized transport are the only bases of other expressions of tissue polarity. The localization of new shoots is generally much less clear than that of the roots. This localization could be an expression of the early drainage of auxin from the apical parts. Polar localization of new shoots could also depend on a correlative advantage of the shoots that morphologically overtop all others. These possibilities require further study.

It thus appears that Vöchting's observations concerning organ regeneration are, as he claimed, expressions of an innate tissue polarity based on each cell acting as a minute 'magnet', transporting morphogenetic signals in one preferred direction. The polarity of plant regeneration need not require any 'prepattern': the cells can all be identical. In contrast to plants, polarity of regeneration in animals such as *Hydra* depends on gradients of signals or of tissue competence (Bode and Bode, 1984). The difference between plants and animals may be related to the distance over which development occurs. Gradients over the distances characteristic of plants may not be steep enough to be reliable. The action of polar transport, on the other hand, need not be diminished even over very large distances.

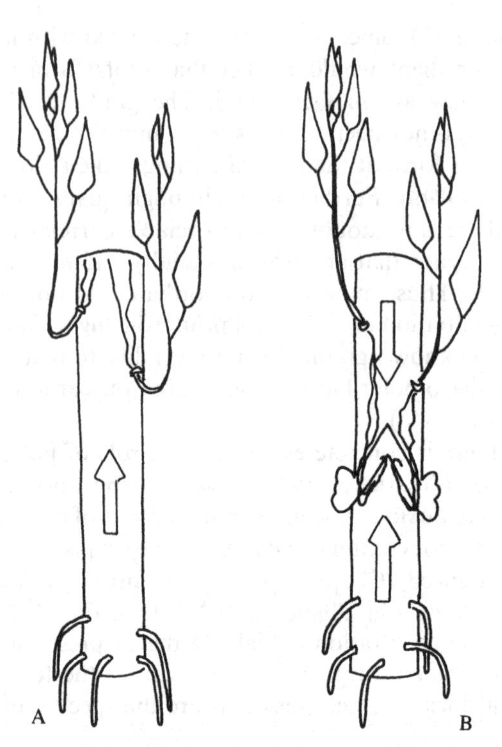

Figure 5.4. Two types of 'inverted cuttings'. Arrows within the cuttings mark the original polarity, the direction to the roots in the original intact plants. A. Roots and shoots were induced in unusual locations by local environmental conditions (light or darkness, high humidity and gravity) and by the removal of unsuitable organs when they first appeared. Such cuttings live for some time, but tissue polarity prevents the formation of vascular contacts between the new shoots and the new roots and the cuttings do not survive. B. Inverted grafts: cuttings on which roots and shoots had already formed were grafted so that their polarities were opposed. Such grafts join, though a large mass of callus appears at the graft region. The new vascular tissues induced by the growing buds follow tissue polarity – and fail to form new shoot-to-root contacts. Thus, the grafts join but the plants eventually die.

5B. THE INDUCTION AND STABILITY OF TISSUE POLARITY

Following his conclusions about the polarity of plant tissues, Vöchting (1892) also attempted to reverse this polarity in various ways. He grew plants in which both shoots and roots were present but a part of the axis between them was inverted (Fig. 5.4). This inversion of the original shoot–root orientation was achieved either by grafts of inverted tissue or by removing new organs so that only shoots placed morphologically below the roots were able to develop. Such plants with an inverted axis survived for some time and their grafts, where present, joined (Chapter 9). Yet sooner or later most of the plants in which part of the axis had been inverted died. Vöchting (1892) considered this as evidence that the polarity of plant tissues is stable – it cannot be inverted even when inversion is required to save the life of the plant.

Vöchting's observations went further. He noticed that new vascular tissues were formed in the plants with inverted shoot – root relations. The orientation of these new vascular contacts corresponded to the original rather than the actual locations of the shoots and the roots (Fig. 5.4). Thus, vascular differentiation in these plants expressed an innate orientation or polarity of the tissues rather then the functional needs of the plant. This aberrant orientation meant that new functional contacts were not formed and this could well have been the cause of the eventual death of the plants.

These observations on vascular contacts mean that polarity can be seen microscopically, at a cellular level. Unlike Vöchting's conclusions concerning the polarity of organ regeneration, the observations of the polarity of vascular differentiation have been generally misunderstood or forgotten. Yet evidence considered in Chapters 2 and 3 would suggest that the same tissue polarity is involved: vascular differentiation is causally related to the development of organs and an important signal for vascular differentiation is auxin, the hormonal signal of the shoot which also regulates root initiation.

INDUCTION OF VASCULAR DIFFERENTIATION BY THE FLOW OF AUXIN

The polarity of vascular differentiation is evident in intact plants (Sachs, 1981a). In any given location differentiation has a clear axis, along vascular strands and their transporting channels. A direction, or polarity, along these strands can generally be defined (Chapter 5) since they connect two different centers of development, shoot and root apices. This raises the general question of what controls the location and orientation of new vascular differentiation, a developmental process which, in plants with a cambium, extends all the way between the shoots and the roots.

Both shoot and root apices induce the differentiation of vascular tissues connecting these apices with the rest of the plant (Chapter 2). This in itself would not explain how the vascular tissues induced in the shoots and the roots form one continuous system. But shoots and roots have complementary rather than identical effects on vascular differentiation (Fig. 5.5; Sachs, 1968). Developing shoot tissues actively induce vascular differentiation and one of the signals of this induction is auxin (Chapters 2 and 3). Developing root tissues orient differentiation towards themselves,

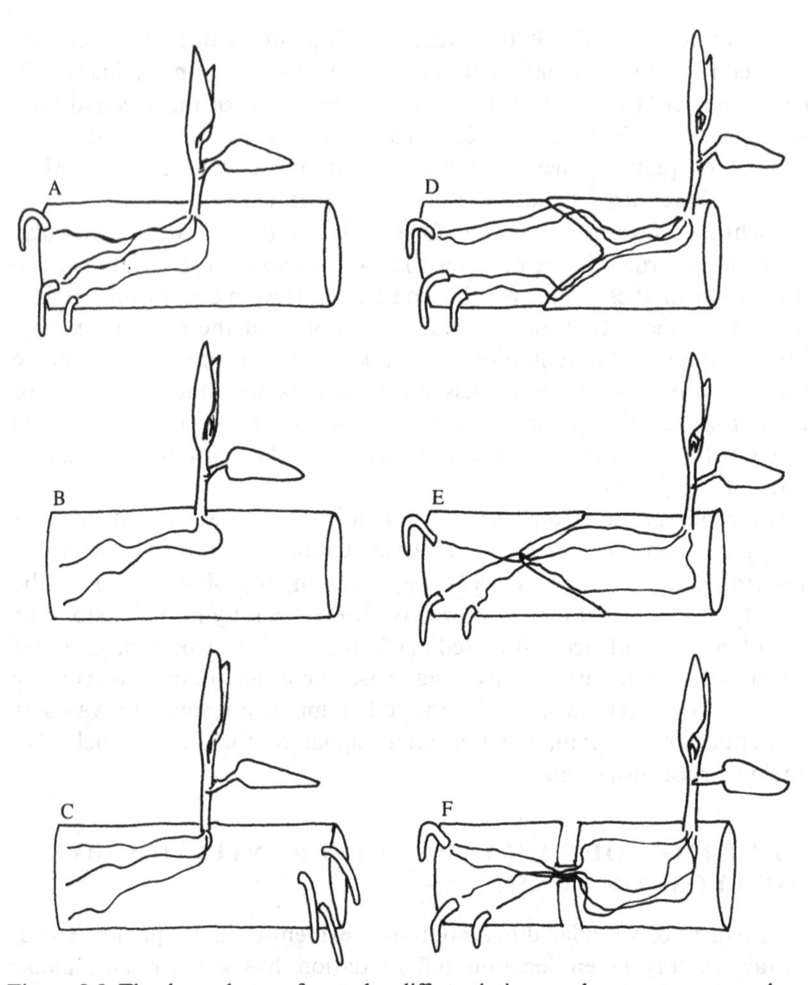

Figure 5.5. The dependence of vascular differentiation on shoots, on roots and on tissue polarity. A. In a regenerating cutting, vascular differentiation connects a new shoot with the new roots along the original tissue polarity. B. Vascular differentiation may occur even when the tissues are not competent to form new roots. The location of the vascular differentiation is determined by the development of the new shoots. C. Roots induced by local environmental conditions in unusual locations do not reorient vascular differentiation and do not cause it to divert from the course dictated by polarity. D, E. Vascular contacts are formed across grafts. The influence of the two graft members is not symmetrical: vascular differentiation occurs on the shoot side of the graft as much as possible. These asymmetries could be expressions of the influence of tissue polarity on vascular differentiation. F. Vascular tissues regenerate around wounds and the regenerated strands have the form of a flow system. Differentiation occurs closer to the wound on the shoot side, where it is 'pushed' by the polar transport of auxin.

but new vascular tissues can be formed in the absence of any roots. Such vascular differentiation occurs in flaps of stem tissue even if they do not lead to any existing roots and never initiate new roots (Fig. 5.5). The orienting effects of roots on vascular differentiation cannot be replaced by any known substances not even by the cytokinins that serve as root hormones (Chapter 3; Sachs, 1981a). Of course, in the long run removing the roots stops differentiation, but this could well be indirect and depend on the general stoppage of all development – including the development of shoot apices which induces vascular differentiation.

Only one mechanism is compatible with the known effects of roots, with their orienting vascular differentiation whenever they are present, without their being essential for the actual occurrence of this differentiation. This is that roots act as preferred *sinks*, and thus direct into themselves the differentiation-inducing signals that originate in the shoot tissues (Sachs, 1968, 1981a). If this is true then vascular differentiation would occur along the flow of inductive signals, from the developing tissues of the shoots to the roots [Sachs, 1969a, 1981a). This would account for the coordination of the effects of shoots and roots and for the differentiation of vascular tissues throughout the plant.

The response to signal flow would be at the level of the entire tissue; the individual cells could be sensitive to the local gradients associated with this flow or to the actual flow of signals through the cells. Controls at the cellular level are not stressed in the present discussion, but there is some evidence favoring the possibility that cells are sensitive to actual flow (Sachs 1981a; Sachs & Cohen, 1982). On the other hand, the concept of a determination of a differentiation pattern by flow through a tissue is central to the entire discussion of polarity. It is the basis of hypotheses of polarity induction and the patterning of vascular differentiation (below and following chapter) and, to the extent that these hypotheses correlate and account for many facts, they support the concept of signal flow. The following points are additional evidence for differentiation being dependent on signal flow:

(a) In plants that have a cambium, wounds are followed by a regeneration of the longitudinal continuity of the vascular system (Fig. 5.5). This regeneration depends, at least partly, on signals originating in the young parts of the shoots (Jacobs, 1952, 1970, 1979). These facts can be readily understood if there is a continuous flow of differentiation-inducing signals even in intact plants. Regeneration would be an observable expression of signal flow as it becomes re-established around wounds. Local effects of the wound, independent of its interruption of flow, need not be important: differentiation occurs only when the continuity of the vascular system is interrupted. Furthermore, regeneration establishes continuity between vascular strands and it need not follow the contours of the wound.

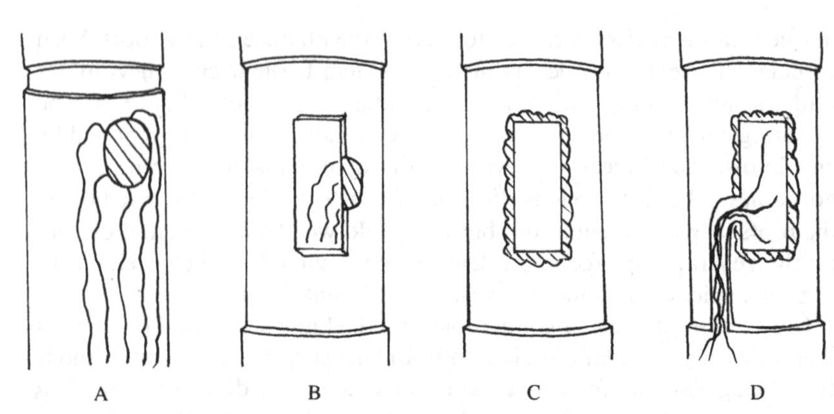

Figure 5.6. The relation of polar vascular differentiation to auxin flow. A. Vascular differentiation associated with a local source of auxin. The effects of such auxin were most pronounced below a girdle, as shown in the figure. The new vessels are all oriented by the auxin source and they follow the polarity of the tissue, towards the roots. B. Local auxin has an inductive and orienting effect even when it is applied to a small region that is completely girdled and is connected with the rest of the plant only through the mature xylem. C. There is no visible induction when auxin is applied on all sides of a completely girdled region: when no flow of auxin could be expected. D. A similar girdled region flooded with auxin but connected to the rest of the plant by a narrow bridge of cambial and other tissues. Differentiation marks the course of auxin flow towards the roots.

(b) The general form of the vascular tissues in intact plants resembles a drainage system connecting the shoots with the roots. This system is polar: individual vessels or sieve tubes connect shoots with roots and not organs of the same type (see following chapter). There are, rarely, individual vascular elements that are isolated and do not appear to be part of any flow system. These may reflect a local flow (Sachs, 1981a; Burgess and Linstead, 1984). They may also be parts of strands that, except for isolated cells, did not reach the stage of final differentiation (Sachs, 1981a). The resemblance of the vascular tissues to a flow system is clearly seen in the pattern of their regeneration. This is obvious where only a narrow bridge of tissue is left to connect the shoots with the roots (Fig. 5.5F). Here, too, the pattern on the two sides of wounds is not symmetrical: it fans out gradually only on the root side (Sachs, 1981a).

(c) Local application of auxin to plant tissues that are either embryonic or close to a wound causes vascular differentiation (Jost, 1942; Jacobs, 1952; Wetmore and Rier, 1963; Sachs, 1981a). These are conditions in which a gradient or flow of the auxin in the tissue could be expected. Merely raising the concentration of auxin by applying it from all directions to small regions of tissue, ones that do not develop internal sinks for auxin, does not result in vascular differentiation (Fig. 5.6c; Sachs, 1981a).

Objections to the concept of induction by a flow of signals from the developing shoot tissues to the roots have been based on observations of the course of undisturbed vascular differentiation (Esau, 1965; Larson, 1975). Mature cells are often seen at the base of a shoot apex and maturation proceeds gradually towards, rather than away from, the tip of the shoot. Various other patterns of maturation have been observed, such as the maturation of xylem first at the base of a leaf, this maturation later proceeding both into the leaf and basipetally along the stem. In large trees, vessels may appear first near developing buds and the wave of maturation proceeds basipetally as though it is an observable expression of an inductive flow of auxin (Tepper and Hollis, 1967; Lachaud and Bonnemain, 1982). But vessel maturation may also follow various other patterns, and these patterns may be quite complex. The important point is that what is observed is the sequence of maturation, and not a 'wave of induction'. Maturation depends primarily on local events, on the 'physiological age' of the tissue, and not on small differences in the time of induction. For example, early vessel maturation at the base of a leaf could reflect the limited elongation of this region relative to the tissues on both of its sides, not some local origin of vessel induction (Sachs, 1981a). It is quite possible, furthermore, that both controls that operate towards and away from the young shoot tissues exist, and their relative role depends on the plant, tissue and growing conditions. The important conclusion for the discussion below is that there is no evidence that conflicts with the concept that vascular differentiation is induced by a flow of signals moving away from the young shoot tissues, even though there may be additional determinants of this differentiation.

EVIDENCE OF POLARITY CHANGES

It was suggested above that vascular regeneration depends on the same control that determines differentiation in intact plant. It is thus important that vascular regeneration occurs at *all angles to the original shoot–root axis* (Fig. 5.7; Janse, 1914; Neeff, 1914; Kirschner et al., 1971; Sachs, 1981b). If this regeneration reflects tissue polarity it must mean that a new polarity can be induced even in relatively mature regions of the plant. Since regeneration around wounds occurs over considerable distances, measured in millimeters in herbaceous plants and in centimeters in trees, it is relatively amenable to experimental manipulations.

The cells of the regenerated tissues do have distinctive forms: they often have rectangular shapes and if they are elongated it is along the original axis of the tissue and not the axis of the new strand (Fig. 6.1B; Kirschner et al., 1971). But vessel and sieve tube elements of the regenerative type also occur in plants that have not been wounded in any way: they are common at the base of lateral roots and shoots, especially when these develop on regions that are relatively mature. These non-embryonic

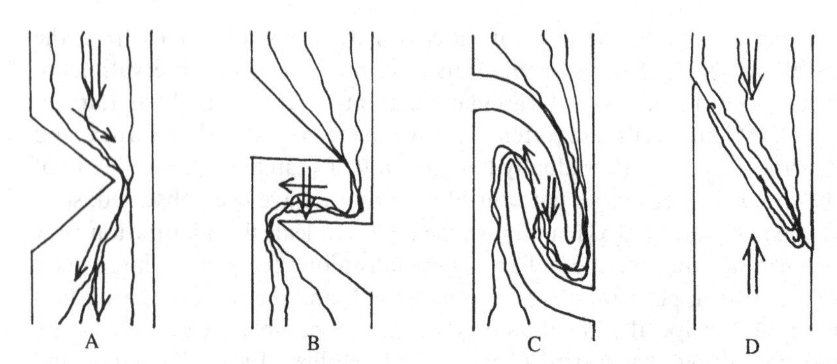

Figure 5.7. Polarity changes expressed by vascular orientation: apparently contradictory results indicating that the polarity of the very same tissues may be either labile or determined, depending on the experiment. The experiments were performed on bean hypocotyls; wavy lines indicate new vessels and the original vessels are not shown. Arrows indicate polarity: the ones with double lines show the original polarity of the tissues and the ones with single lines show the new polarity indicated by new vessels. A. Regeneration around a severe wound. The vessels formed above and, to a lesser extent, below the wound are at various angles to the original longitudinal axis of the tissue. B. Two wounds cut all the original, polar connections, leaving only a horizontal bridge between the shoot and the root. Regeneration at right angles to tissue polarity occurs readily, within 3 days. C. As in (B), but the wounds left a bridge in which regeneration could only occur opposite the original shoot-to-root polarity. Such vascular regeneration took place provided the 'inverted' bridge was no longer than a few millimeters. D. A graft in which the upper and the lower members had opposite polarities. In contrast to the various wounds shown in (A)–(C), these grafts demonstrate a determination of tissue polarity: new vessels stay close to the original shoot–root axis. No functional vascular contacts are formed and the plants eventually die.

tissues respond to differentiation signals after the general shape of the cells has been determined (Kirschner et al., 1971). It is the nature of the differentiating tissues, not differences in the controlling signals, that is the cause of the special traits of regenerative vascular elements.

If the regeneration of vascular tissues reflects a new polarity, there should also be a new polarity of auxin transport. This would be a new trait of the tissues, not a temporary expression imposed by external conditions (Sachs, 1984a; Wright, 1981). Experimental evidence concerning this important possibility is meager, partly because regeneration along new polarities occurs over distances that are not convenient for the study of auxin transport. *Tagetes* plants have an exceptional ability to develop even when considerable lengths of their stem have an 'inverted polarity' (Fig. 5.4). Measurements by Went (1941) indicated that stems of such inverted plants have a new polarity of auxin transport. However, Sheldrake (1974) could not repeat Went's results. It is possible that a new polarity of transport is induced, but only in the cambium and its immediate products. Thus, a new polarity of auxin transport would be masked by the effects of the unchanged tissues.

The problem of recognizing a new auxin polarity without its being masked by original traits was solved by following transport where there was a 90° change of tissue orientation (Gersani and Sachs, 1984). Since there is no appreciable polarity at right angles to the plant axis, this new 90° polarity would appear over a clear background. Gersani and Sachs (1984) studied the transport characteristics of transverse tissue bridges of bean hypocotyls (as in Fig. 5.7B). These bridges were cut from the plants after regeneration had occurred, and their auxin transport was studied. A preferred transport of radioactivity applied as auxin was found along the new shoot-to-root direction. Although additional studies are called for, this result shows that vascular regeneration is an indication of a new transport polarity.

Regenerative vascular tissues, regardless of their orientation, perform transport functions. This is shown by the ability of plants that had been severely wounded to resume normal growth and, as expected, this resumption of growth is correlated with the appearance of new vascular tissues. More direct evidence comes from studies of the transport of radioactive substances (Gersani, 1985). Double wounds that cut all vascular tissues (Fig. 5.7B) were found to interrupt all measurable transport of sucrose. The movement of radioactivity applied as auxin preceded both overt vascular differentiation and the movement of radioactive sucrose. These results are in accordance with the hypothesis that the resumption of auxin flow along new polarities is an early stage of the regeneration of the entire vascular system.

A COMMON BASIS FOR POLARITY REORIENTATION AND POLARITY DETERMINATION

Vascular differentiation was used above to demonstrate two traits of polarity that are in apparent mutual contradiction (Fig. 5.8). On the one hand, the polarity expressed by vascular differentiation was stable or determinate in Vöchting's and other, more recent, experiments (Fig. 5.8 A, C–F; Thair and Steeves, 1976). But the very same polarity of vascular differentiation can be readily changed by wounds (Fig. 5.8B, E; Janse, 1914; Neeff, 1914). Both these general results can be readily repeated with bean hypocotyls (Fig. 5.7; Sachs, 1981b). There must therefore be differences in the conditions in which polarity is either maintained or changed.

A survey of many observations and literature reports leads to a generalization (Sachs, 1981b) which is illustrated by the experiments shown in Fig. 5.8. Differentiation follows existing polarity whenever possible. This is true even when the new vascular tissues serve no

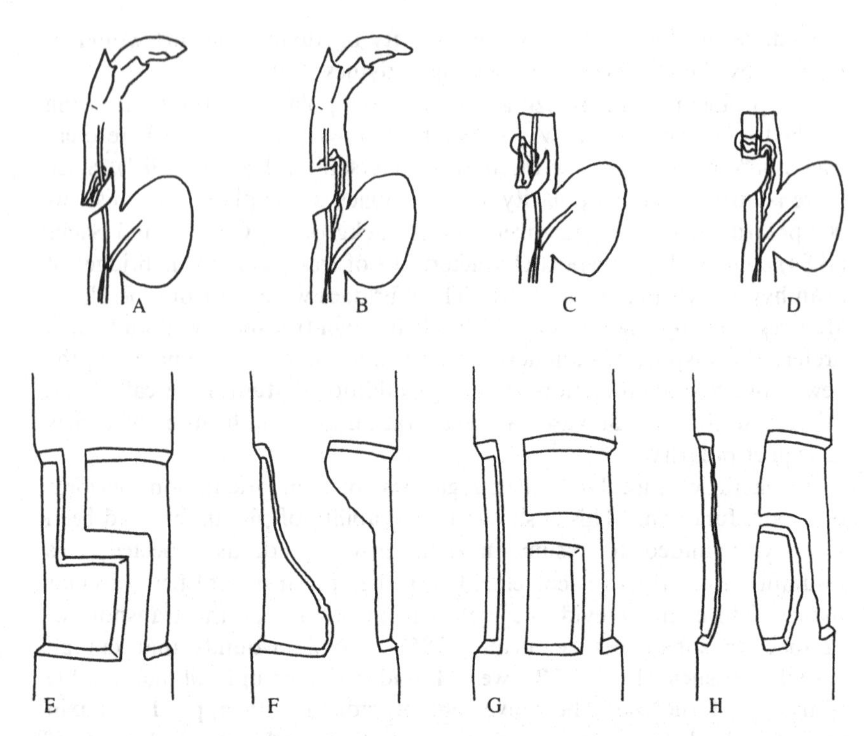

Figure 5.8. New vascular differentiation where there are alternatives of different polarities. The experiments show that reorientation occurs readily only when there is no alternative course, one which maintains the original polarity. (A)–(D). Wounded pea seedlings. A. New differentiation proceeds into a flap of tissue and new contacts with the roots are not formed. If the flap of tissue is of sufficient length – a few millimeters – the shoot often dies. B. A similar pea seedling in which an additional cut removed the flap of tissue. In these plants new contacts with the roots are always formed, even though they require a reorientation of the original tissue polarity. (C), (D). The same results as in (A) and (B) were obtained when vascular differentiation was induced by an exogenous source of auxin rather than by a young shoot. (E), (F). Early and late stages of a bridge of cambium left across a wide girdle of a branch of an *Ailanthus* tree. Although the bridge included a horizontal region, cambial activity was always resumed and it formed new vascular contacts with the roots. Within a few months the bridge became relatively straight. (G), (H). As in (E) and (F), but an additional, direct contact was left as an alternative to cambial reorientation. Most cambial activity occurred through the direct alternative; the transverse bridge did not develop and it often withered and died. (Based on Kirschner and Sachs, 1972).

functional purpose and the shoot may die for lack of regenerative contacts with the cotyledons and roots. On the other hand, when no polar alternative is available differentiation along a new polarity occurs readily. Even a complete reversal of the original polarity is possible along distances of a few millimeters (Fig. 5.7C).

These facts offer an explanation of the apparent contradiction concerning the stability of polarity. Differentiation is always associated with auxin flow. Whenever possible, this flow is along the original polarity of the tissues – since polar transport is far more rapid and efficient than diffusion, it is the preferred alternative. The other possibility is that the flow is determined by diffusion (Sachs, 1981a, b). This diffusion is towards the closest sink for auxin: this is usually some polar tissue, preferably vascular strands, which lead in the general direction of the roots. The actual presence of roots is not essential, because polar transport continues to remove auxin even in their absence. If no such sink is available, whirlpools and closed rings of vascular tissues are formed (Sachs and Cohen, 1982).

It is now possible to offer a general hypothesis. Polarity is both induced and expressed by the oriented flow of auxin. Induction is a gradual process, the polarity of the tissues increasing as the flow of auxin continues. These generalizations would mean that there is *a positive feedback relation between polarity and auxin flow* (Sachs, 1981b). The positive feedback could be the basis of the determinate nature of tissue polarity: this polarity always prevents auxin diffusion, and diffusion is the only way a new polarity can be induced. It also accounts for the rapid induction of a new polarity, expressed by vascular differentiation, when no polar alternative is available – for then the positive feedback that confers stability is disrupted.

THE CELLULAR BASIS OF TISSUE POLARITY

An important characteristic of the controls considered here is that they specify the *orientation* of events and not only the differentiation of cells. The flow and/or the gradients of auxin have been shown to have a unique role in the specification of this orientation (Chapter 3; Czaja, 1935; Ruge, 1937; Balatinez and Farrar, 1966; Sachs, 1981a, b). Though the central topic above and in the rest of this book centers around tissue organization, at least a brief digression to polarity at a cellular level is required by the discussion of polarity determination.

The polarity of individual cells could not be due only to differential gene activity: as far as is known at present, such activity could not have any orientation. Since the cells are polarized, something other than the genes or chromosomes must be oriented or localized within them. Various possibilities may be classed in three groups: gradients of substances and even organelles within the cells; oriented aspects of the cytoskeleton; and local differences in cell membranes (for additional discussion of oriented cellular interactions see Chapter 9). Observations of living cells during changes of polarity leading to vascular regeneration (Kirschner and Sachs, 1978; Schulz, 1988) show that re-orientation can be recognized in the cytoplasm and that it involves various aspects of the cell structures. This means that the question must be not only which structures are involved in polarization but also whether an initial, controlling event can be recognized. For the sake of clarity the various possibilities will be taken up separately.

Cytoplasmic gradients within cells have been recorded in many different systems (Bünning, 1957; Burgess, 1985; Schnepf, 1986). These gradients are expressed by the localization of the nucleus, by the concentrations of the cytoplasm and by the local densities of almost all cytoplasmic organelles. Such gradients have been shown to be early stages of various polar events. For example, cytoplasmic gradients precede the unequal divisions that lead to the spacing of stomata and other idioblasts (Chapter 8). But the polar phenomena considered in the present chapter occur over large distances and do not necessarily originate in any single cell. Furthermore, at least at present there is no evidence that the polarity of auxin transport is due to the unequal distribution of cytoplasm or of the organelles within the cytoplasm.

The best known and largest component of the cytoskeleton are the microtubules. The distribution of these is certainly related to other aspects of polarity (Hardham and McCully, 1982; Gunning and Hardham, 1982), especially to the plate of cell divisions and to the orientation of the microfibrils in the cell walls. Microtubules are the most likely skeletal structure of the oriented cytoplasm strands to be associated with tissue polarity (Kirschner and Sachs, 1978; Schulz, 1988). There is evidence that microtubules can be re-oriented by auxin sources (Bergfeld et al., 1988). There is perhaps more evidence that the orientation of the microtubules is influenced by ethylene (Lang et al., 1982; Roberts et al., 1985) and that ethylene disrupts various aspects of tissue polarity (Bünning & Ilg, 1954). It is hardly likely that ethylene, being a gas, supplies the cells with directional information. A reasonable hypothesis concerning directional effects of ethylene is that it changes the polarity, and hence the polarizing effects, of auxin transport (Morgan and Gausman, 1966). All this shows that microtubule orientation is tightly linked to tissue polarity; but there is also evidence that the microtubules are not the initial control of orientation. Colchicine is known to disrupt microtubule organization, and, as expected, colchicine treatments result in the formation of vessels with disorganized, disrupted cell walls (Hepler and Fosket, 1971; Hammersley and McCully, 1980; Burgess, 1985). The critical fact is that even in these conditions, where colchicine is clearly acting on the cells, new vessels are formed with the member cells arranged in distinct files (Sachs, 1983).

The available evidence concerning the polar transport of auxin suggests that it depends on a localization of specific proteins, channels for auxin, at the basal side of the transporting cells (Rubery, 1987). This would mean that difference in the orientation and in the degree of tissue polarity are a function of differences in the number and the degree of the localization of these macromolecules. An interesting attempt to visualize these differences by using fluorescent antibodies (Jacobs and Gilbert, 1983; Jacobs and Short, 1986) has not yet been confirmed and extended. Using a light microscope it is difficult to distinguish between macromolecules in the lower part of one cell and in the upper part of its neighbor, but the discussion above suggests a critical test for such staining procedures: the molecules should not only appear on one side of cells but their location should also change when polarity is reoriented by wounds or by auxin treatments (Figs. 5.7B, 5.8D; Gersani and Sachs, 1984).

The suggestion to emerge from all this is that polarization involves many aspects of cell structure, the most likely initial events being differences in membrane characteristics, possibly the localization of protein channels. Perhaps this is a principle that is much more general, involving polarization events in additional organisms and signals other than auxin. An early determining event in the polarization in zygotes of brown algae is the appearance of an oriented electrical current (Jaffe and Nuccitelli, 1977; Jaffe, 1981). It has been suggested that this current is both localized by and further localizes specific channel proteins at one pole of the cell membranes. Measurements have since shown that electrical currents are characteristic of polarized growth in plant systems (Weisenseel et al., 1979). Currents are also characteristic of the early stages during the organization of new embryos in tissue cultures (Brawley et al., 1984; Gorst et al., 1987; Rathore et al., 1988). The same principles of an oriented 'flow' which reinforces itself and leads to overt differentiation appear in relation to both electrical currents and auxin transport. It is an intriguing possibility that auxin transport, which occurs even in embryos (Fry and Wangermann, 1976) and appears to play a role in their organization (Schiavone, 1988), is but a late manifestation of an earlier, less-specific polarity expressed by electrical currents.

6

The canalization of vascular differentiation

It was concluded in previous chapters that vascular differentiation is induced by young shoot tissues and by localized sources of auxin. Furthermore, this differentiation is oriented along tissue polarity that determines auxin transport. It is now possible to turn to cellular patterns within the vascular strands. Vessels and sieve tubes, the transporting channels of the vascular system, differentiate as long files of similar cells

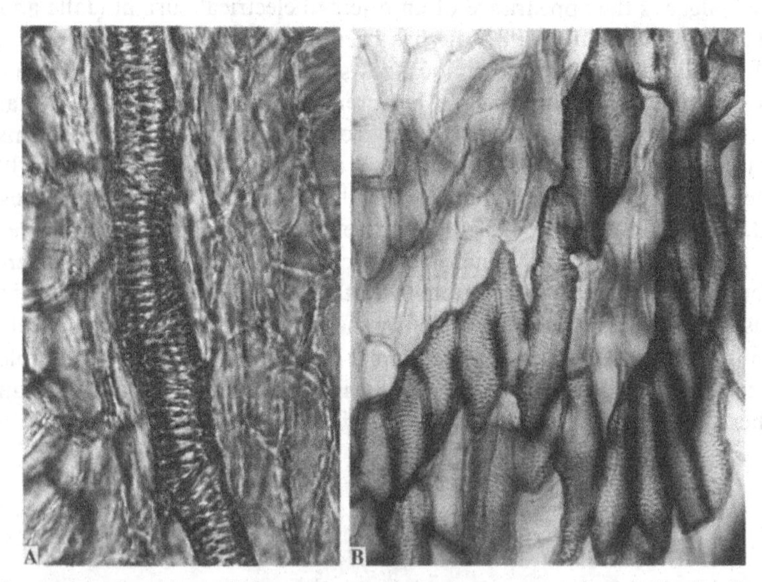

Figure 6.1. A major cellular pattern: the arrangement of specialized cells in longitudinal files. Vessels consist of many cells which are connected by specialized holes. These cells are embedded in a matrix of other cells, not readily seen because of their thin walls. A. A normal vessel in the storage tissue of a turnip (*Brassica napus*) root. A thick section cleared with lactic acid. X 150. B. New vessels induced in turnip storage root by local application of auxin (1 % indoleacetic acid in lanolin). The flow of this auxin, and the resulting differentiation, was at right angles to the shoot–root polarity of the tissues. Files of specialized cells are evident even when their axes do not correspond to the original longitudinal axis of the tissue. X 200.

(Fig. 6.1; Essau, 1977; Fahn, 1982). There are distinctive structures – openings in the cell walls – between cells along a file. The neighbouring cells of the transporting files differentiate in other ways, unless they form part of a neighboring transporting channel.

Vascular differentiation is thus an example of a major cellular pattern, and this is a pattern that can be induced by known hormonal signals (Chapters 3–5). A general problem arises: how do inductive effects become canalized so that only distinct files of cells differentiate in special ways. Within vascular strands, furthermore, various cellular types develop in predictable though complex patterns. These problems of canalized differentiation of files of cells and of complex vascular strands are the topic of the present chapter.

THE CANALIZATION OF SIGNAL FLOW

A local source of auxin induces the differentiation of both organized vessels and of sieve tubes (Fig. 6.2; Jacobs, 1970; Sachs, 1981a). The pattern of specialized cell files must therefore depend on information and events in the differentiating tissue itself: it could not be specified merely by the exogenous source of the simple inductive molecule, auxin. The special files could either be present in the tissue before it is induced to differentiate or they could be formed during the process of differentiation itself. An early presence of the pattern would be a 'pre-pattern' of cellular properties that are not necessarily observable (Chapter 1), and in shoot and root apices such properties could even be localized in single cells, these becoming long files when the tissue grows. The other possibility, the formation of the cellular pattern during the process of differentiation, requires two types of intercellular correlations: an induction of the differentiation of similar cells along the future transporting channel and inhibitory effects along the transverse axes, reducing similar differentiation.

Evidence relevant to the choice between these two possibilities can be found in the regeneration of vascular tissues (Chapter 5). New strands can be induced by newly growing buds, grafted shoots, wounds and local applications of auxin (Sachs, 1981a). When a tissue responds to these treatments the precise location of the new strands, and even their orientation, can be varied at will (Fig. 6.2). This leaves no doubt that the competence to differentiate could not have been limited to strands of special cells. It can be concluded that at least regenerative vascular tissues do not reflect only a pre-pattern of cellular properties. It is difficult to make the same clear statement concerning events within intact meristems. However, all types of vascular tissues appear to be under the same controls (Sachs, 1981a). The assumption of a pre-pattern that precedes vascular differentiation in meristems, furthermore, is contradicted by the

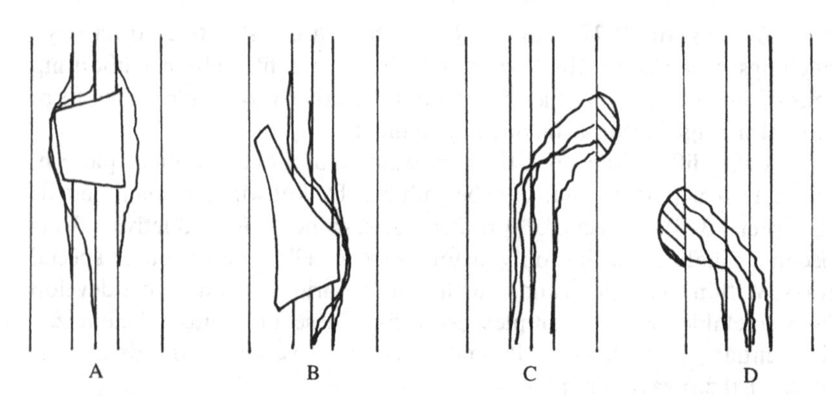

Figure 6.2. The location of new xylem induced by wounds and by auxin does not depend on previous cellular patterns or pre-patterns. A, B. Regenerative vascular tissues (wavy lines) depend on the chance location and the form of wounds. C, D. Local auxin (1 % indoleacetic acid in lanolin) application to pea (*Pisium sativum*) epicotyls caused the differentiation of vascular strands. These strands could form in the entire tissue – their location in any given case was determined only by the applied auxin.

evidence of chimeras (Chapter 7) concerning the varied origin of vascular channels.

Thus a central problem concerning the control of the pattern of vascular channels is why inductive effects influence only certain files among cells that are potentially equal. This appears to be a difficult question, about which not much information is likely to be available. But the discussion of polarity (Chapter 5) offers a factual basis for considering this cellular patterning. It was concluded that vascular differentiation depends on a polar flow of auxin through the cells. The canalization of differentiation reflects, therefore, a canalization of the flow of auxin and presumably other, unknown signals. The question thus becomes one of how the flow of specific substances is canalized to files of cells, leaving their neighbors to differentiate in other ways.

Vascular differentiation and polar auxin flow enhance one another: the differentiation that is induced along the axis of flow leads to an increased capacity for transport along the same axis (Chapter 5). This suggests an hypothesis as to how differentiation and transport could be gradually canalized (Sachs, 1969a, 1981a, 1984a, 1986; Mitchison, 1981). It could be expected that *an inductive flow which at first influences many cells is carried by progressively fewer files as the cells differentiate and become better transporters*. Thus, the same processes that orient differentiation (previous chapter) could account, if they occurred gradually, for the canalization of both differentiation and transport – and for a major pattern of specialized cells (Fig. 6.3). The critical cellular process required for this patterning would be a gradual, inducible increase in transport capacity of the cells.

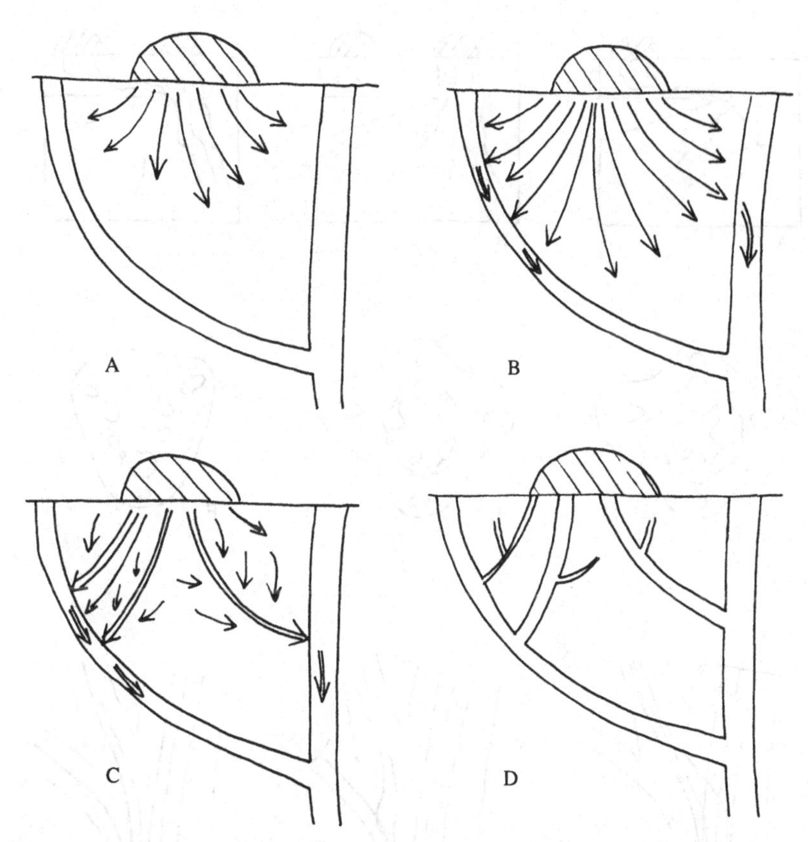

Figure 6.3. Schematic representation of the canalization hypothesis concerning the induction of strands of specialized vascular cells. A. Auxin diffuses in all directions from an exogenous source. B. The diffusing auxin reaches vascular strands, where it is removed rapidly. As a result of this removal, flow continues towards the strands much more rapidly than in all other directions. C. The cells through which auxin flow continued to specialize, becoming preferred channels for the axial transport of auxin. These cells thus drain the auxin in their vicinity, canalizing both the flow of auxin and the continued differentiation which this auxin induces. D. The cells through which auxin flow has continued for a sufficient length of time differentiate as mature vascular channels. They thus become the sinks for additional, lateral auxin flow.

Evidence supporting the canalization hypothesis

A major advantage of this hypothesis is that it can be tested in various ways. It has been subjected to mathematical analysis (Mitchison, 1980, 1981): It was shown that the intuitive suggestion of a canalization resulting from a positive feedback between signal flow and transport capacity could be correct for reasonable parameters of permeability and flow. The hypothesis is based on the experimental evidence considered in

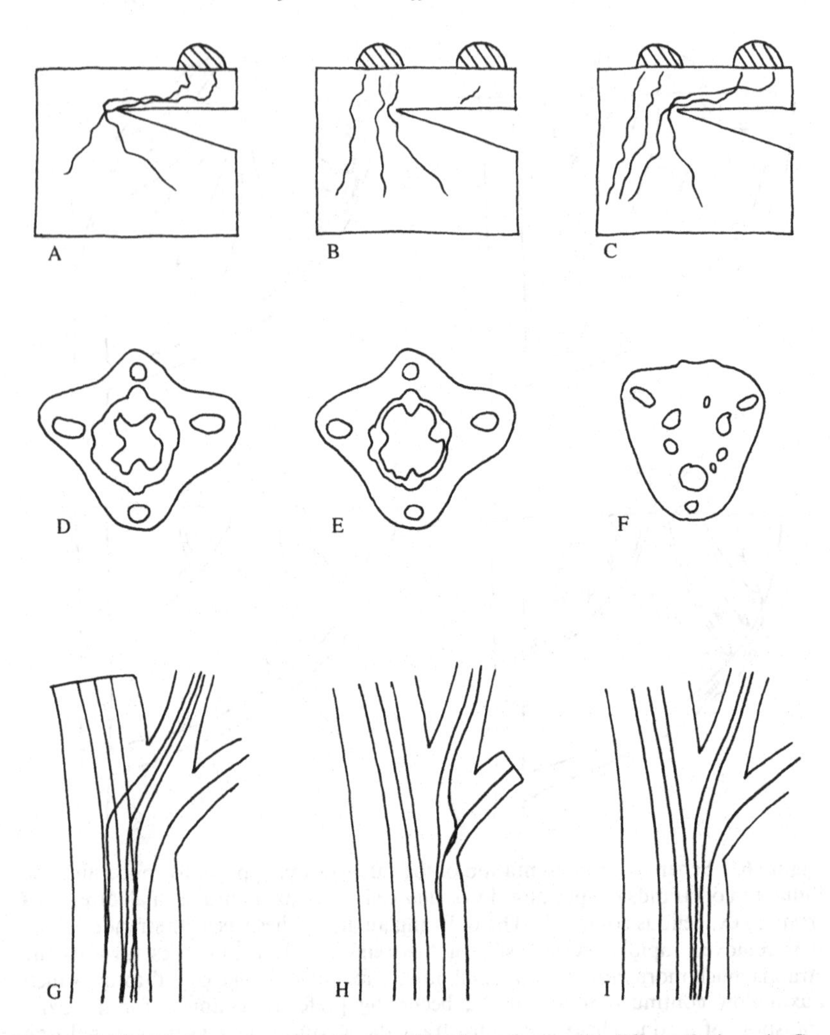

Figure 6.4. Evidence concerning the canalization of vascular differentiation. A–C. An experiment demonstrating that vascular induction is a gradual process, requiring the continued flow of auxin. Pieces of turnip storage tissue were cut as shown and auxin was applied locally. A single source of auxin (A) always induced strands around the wound. When two source were applied together (B) differentiation indicated that the source on the left inhibited the flow of auxin from the source on the right. The important treatments were the application of the two sources at different times. The inductive effects of the source on the right were not prevented (C) only when the application of the source on the left was delayed for 2 days. Thus, the configuration of auxin flow was required for a long period, spanning most if not all of the entire process of vascular differentiation.
D–F. Cross-sections of pea stems showing the inductive effect of young leaves. D. A section taken from an intact control. E. A stem just below a leaf which had been removed when it was in a young but visible stage. Many of the vascular tissues that would have led to this leaf did not mature. F. A stem below a leaf which had been

the previous chapter and evidence that the same feedback principles can extend to the level of files of cells can be summarized as follows (for a more extensive treatment see Sachs, 1981a).

(a) Differentiation is a gradual response. It requires the influence of auxin during most if not all of the time the cells are changing. This conclusion is indicated by the response to temporary treatments with auxin (Roberts, 1960). But the results of such treatments vary, possibly because auxin accumulates in the tissue near the region of application even within short periods of time. The gradual response to auxin can also be demonstrated by experiments in which auxin sources are added rather than removed (Fig. 6,4 A–C; Sachs, 1981b). Such additions can be made at various times after the original application of inductive auxin, and the changed configuration of auxin sources is expected to lead to a change in the pattern of auxin flow. The results of such experiments with turnip (*Brassica napus*) storage roots confirm that vascular differentiation occurs only when there is an unchanged flow for almost two days (Sachs, 1981a,b). This period is long enough for the first appearance of mature vascular elements.

(b) Studies of the transport of radioactive auxin support the canalization hypothesis. Polar auxin transport is possible in most tissues of the plant axis (Goldsmith, 1977). Yet when vascular tissues are present they are the preferred channels of transport, and most transport occurs through them (Sheldrake, 1973; Wangermann, 1977; Bourbouloux and Bonnemain, 1979; Lachaud and Bonnemain, 1984). Radioactive auxin applied to leaves moves rapidly in the phloem, while auxin applied to young primordia that are expected to induce new vascular tissues moves in polar, differentiating cells (Morris and Thomas, 1978; Kaldewey, 1984). Triiodobenzoic acid, known to inhibit polar transport, does not inhibit transport from mature leaves, presumably because this transport occurs with other substances in the sieve tubes (Morris et al., 1973). Lachaud and Bonnemain (1984)

---

removed when it was barely apparent as a bulge on the apex. Not only the vascular tissues are missing but also the fibers and the parenchyma that would normally surround the vascular strands. G–I. The vascular contacts of a growing bud, showing their preferred contacts with cut strands, i.e. those that do not lead to growing organs. G. The main shoot was removed. The bud which was released from apical dominance induced the formation of new vascular strands which joined the cut vascular system of the removed shoot. H. Bud growth was associated with the removal of a leaf. Here, again, the new vascular strands joined the strands that had led to the removed organ. I. A bud that grew on an intact plant, as is common during vigorous vegetative growth or during reproductive development. This bud had independent vascular strands, with no direct contacts with the strands of the neighboring organs.

reported indications for a gradual canalization of the transport of radioactive auxin, from the entire meristem cross-sections close to shoot tips to the vascular regions lower down. This is what would be expected on the basis of the canalization hypothesis, though the result needs to be repeated and extended.

(c) Auxin not only induces new polar channels (Gersani and Sachs, 1984; Chapter 5) but also maintains the capacity of existing tissues for polar auxin transport (Hertel and Flory, 1968; Osborne and Mullins, 1969; Rayle et al., 1969). These results are in accordance with predictions that could be made on the basis of the concept of gradual transport canalization.

(d) The inductive influence of a very young leaf primordium is on the development of a wide section of stem tissue. Only below the apex does the inductive influence become limited, or perhaps canalized, to the vascular system. These statements are indicated by the normal course of shoot development. They are confirmed by experiments (Fig. 6.4 D–F) in which leaf primordia of various sizes were removed from pea shoot apices and the stems which developed were examined (Sachs, 1972a).

(e) As predicted by the hypothesis, vascular strands induce and orient differentiation by acting as sinks for any new flow of auxin. This is shown by regeneration of vascular tissues around wounds and by the contacts between new and existing strands (Fig. 6.4 G–I). New strands, and presumably the auxin flow that induces them, are oriented so that they form contacts with existing strands, but the formation of such contacts with existing strands is inhibited when the strands are connected to young leaves or to supplies of exogenous auxin: the conditions that could be expected to reduce sink activity by flooding the strands with auxin (Sachs, 1969a, 1972a). These simple principles could account for leaf gaps and other characteristics of plant vasculature (Sachs, 1972a, 1981a). The details of the strand patterns, however, may depend on precise, gradual changes in the flow of auxin and presumably other signals, and require further study (Bruck and Paolillo, 1984a,b).

This evidence, coming from different sources, is strong support for the canalization hypothesis. Further support is the ability of the hypothesis to account for a variety of phenomena on the basis of one simple principle. The demonstrated polarization of cells could be the cause of the coordinated differentiation of long files of cells. The same polarization of cells drains the inductive signals away from the neighboring cells that are not part of the same file. Both polar induction and the inhibition of differentiation along the transverse axes could thus depend on the same mechanism and on one signal.

Possible difficulties of the hypothesis

The main published objections to the canalization hypothesis are concerned with the induction of vascular differentiation by signal flow rather than with the patterning of files of specialized cells. These objections were therefore briefly considered in the previous chapter. The most prominent objection is that vascular induction is both upwards and downwards, towards the roots. Induction in both directions, however, is expected if differentiating channels are regions of facilitated flow. Thus differentiation increases inductive flow both away from a source of auxin downwards, along a route of polar transport, and 'upwards' from existing vascular tissues which serve as sinks for the inductive flow.

There are hardly any other hypotheses that deal with the differentiation of vascular elements in defined strands. Though the control of vascular differentiation has received considerable attention (Fosket, 1972; Roberts, 1976; Phillips, 1980; Barnett, 1981), questions of pattern have been generally ignored. Sheldrake and Northcote (1968; see also Sheldrake, 1971) presented evidence that the differentiation of xylem produces auxin. They suggested that this leads to continued induction along the differentiating strand. The auxin they detected, however, could have come from other sources (Sachs, 1981a). It is not clear, furthermore, why they would expect differentiation to be limited to defined strands and why the strands that produce auxin should not get thicker as one proceeds downwards. Another hypothesis is based only on a mathematical model. Meinhardt (1976, 1982, 1984) modified the diffusion–reaction model (Gierer and Meinhardt, 1972) to account for vein patterns in leaves. This suggestion, however, requires not only an inductive substance but also an inhibitor, for which there is no evidence. Induction is supposed to act only from the differentiated cells 'upwards', towards rather than away from leaf lamina tissues. Though in some ways the model is similar to the canalization hypothesis presented above, it is not supported by experimental evidence.

VASCULAR NETWORKS AND THE CONCEPT OF AXIAL CANALIZATION

The canalization hypothesis predicts that vascular differentiation should always be polar: it should connect a source of inductive signals with some tissue that can act as their sink. The prediction is fulfilled for almost all the major strands in the plant, and it is these strands that have been used for experimental work. Though the strands do not have a uniform orientation, they are polar: they connect, directly or indirectly, shoot tissues with root tissues. But a comparative study of the vascular systems of plants does reveal exceptions. These will be considered here under two headings: large

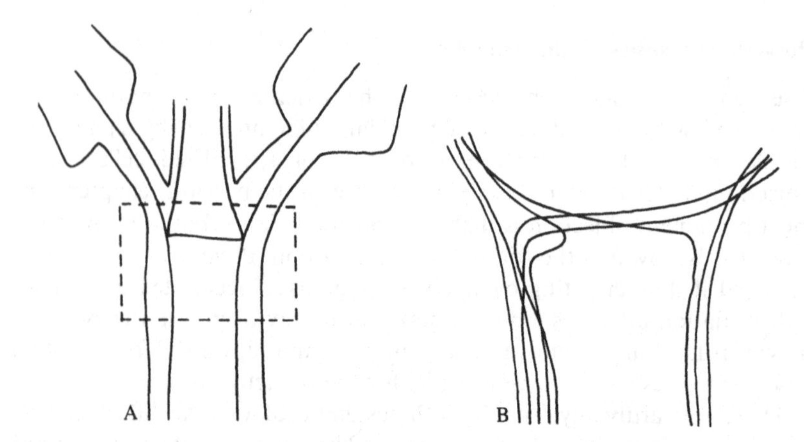

Figure 6.5. The structure of a non-polar strand in bean seedlings (*Phaseolus vulgaris*). A. The first node above the cotyledons is marked by a rectangle. This node has a horizontal strand which connects the 2 opposite leaves. There should be some point along this strand which is physiologically equidistant from the roots and thus has no definable polarity. B. The vessels of the strand within the rectangle in (A). The individual vessels are polar: each connects a leaf with the direction of the roots. Thus, vessels with opposite polarities are included within a single strand. Such strands with no definable polarity could differentiate where the direction of auxin flow changed repeatedly during development.

strands in stems, where details of the structure can be observed, and the vein networks of angiosperm leaves.

Strands with no definable polarity occur in many nodes with two or more leaves (Sachs, 1975a, 1981a). These strands are transverse, connecting vertical strands running between leaves and roots (Fig. 6.5A). Many of these transverse strands connect to the roots on both their sides. It follows that there must be some point along such strands where the physiological distance from the roots is the same in both directions. No signal flow is expected through such non-polar points. If differentiation depends on flow, these strands should not be continuous, and yet even careful observations show no break or change in their vascular continuity.

A way of accounting for the differentiation of these strands with no definable polarity is suggested by observations of the details of their individual vessels and sieve tubes (Fig. 6.5B). Each transporting channel does have a clear polarity: it connects to a leaf on one side and a root on the other. However, transporting channels of opposite polarities are found side by side. If the differentiation of these depends on signal flow, then flow *in opposite directions must occur within the very same strands.*

Observations of the early developmental stages of these strands in bean seedlings suggest, though they do not prove, that the differentiation of vessels with opposite polarities does not occur simultaneously: there might be repeated reversals in the directions of signal flow. An attempt has been made to simulate such reversals in an experimental system (Sachs, 1975a): sources of auxin above a transverse wound in a young pea stem were repeatedly applied and removed. As expected, strands with no defined polarity were induced by such auxin treatments. But these experiments still require refinement and repetition.

These reversals of the polarity of differentiation could result from asynchronous development of the leaves that induce the differentiation of the strand. When one leaf is in a highly active state, its inductive signals would overflow through the transverse strand. They would be directed towards the vertical strands leading to the leaf that is relatively inactive, since these strands would act as a sink for inductive signals. A significant fact is that opposite flows occur within the very same strand. This suggests that the earliest stage of differentiation is a *facilitation of inductive flow along one axis* rather than along one direction or polarity. This is an important modification of the canalization hypothesis. While only the transport of substances of the auxin group is known to depend on a tissue-specific polarity (Chapter 5), an axial preference is a property that has not been studied and it may be relatively widespread (Kaldewey, 1984; Sachs, 1984a).

Axial canalization through single cell files

The second example of strands with no definable polarity is veins in leaf networks (Fig. 6.6). The lack of polarity is evident when arrows are drawn along such veins in the presumed direction of the roots. All such attempts to draw arrows in complex networks of seed plants lead to veins with opposite arrows. On the basis of the large strands considered above, it can be suggested that such veins do result from an inductive flow. This flow, however, would proceed in opposite directions at different times. The additional point is that these veins can include only one vessel and one sieve tube. *Flow in opposite directions would therefore occur within the very same cells.* This could have important implications but it is not, as yet, supported by experimental evidence.

The role of 'axial' rather than polar induction is supported by a comparison of leaves that develop in different ways. Leaves whose primordia develop along their margins, as in many ferns, do not have complex vein networks (Fig. 6.6A; Sachs, 1981a). In most angiosperms the development of the leaf surface, at least as it is seen in stomata maturation (Chapter 8), is not synchronous. If rapidly developing tissues are the main source of inductive signals (Chapter 2), such development could lead to changes in the direction of signal flow. The correlation between complex vein networks and asynchronous leaf development therefore favors the possibility that vascular differentiation is induced first along an axis and only later along a direction or polarity.

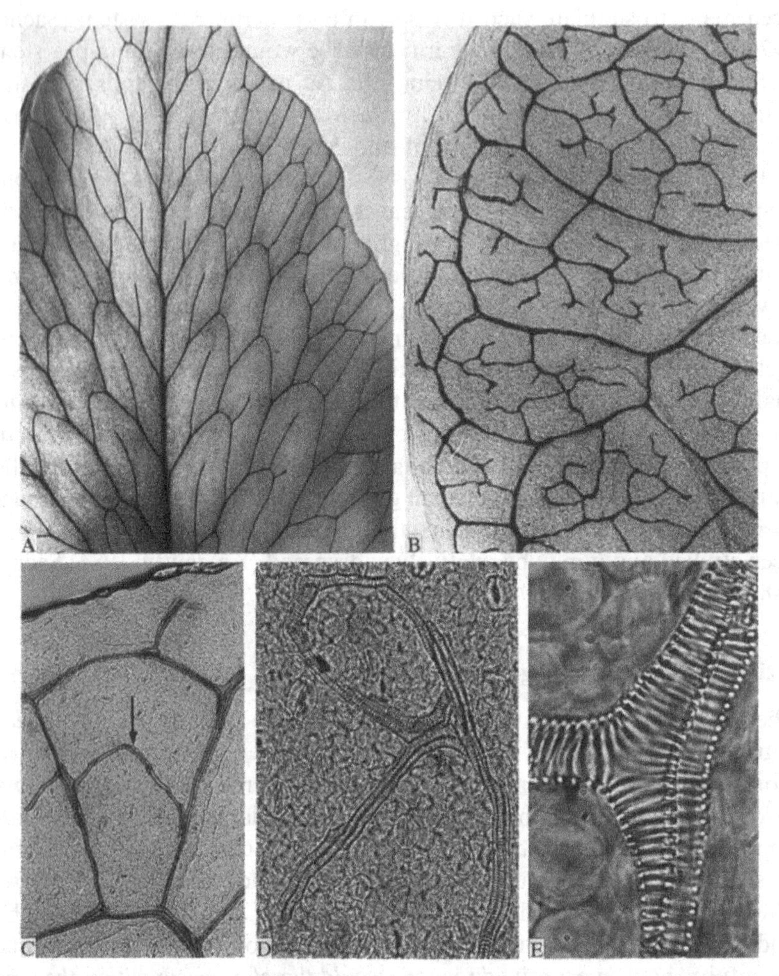

Figure 6.6. Polar and non-polar vein systems of leaves. Photographs of whole mounts cleared with lactic acid. A. The veins of part of a *Cyrtomium* leaflet. Although these veins form a network, it is possible to assign a shoot-root polarity, and thus direction of auxin flow, to all parts of the vascular system. B. Veins of a small region of a pea stipule. The vein network is complex and not readily interpreted as polar. C–E. Details of the xylem in veins of a pea stipule. It is not possible to assign a unique polarity to all parts of these complex systems. In the region marked by the arrow in (C) there must be some point which is physiologically equidistant from the roots on both sides. The round structure in (D) has no obvious direction. The branched vessel membrane in (E) must mean that different polarities occur within one cell. X 2, 35, 10, 30, and 100, respectively.

THE COMPLEXITY OF THE VASCULAR SYSTEM

Facilitated transport of inductive signals deals with the organization of transport channels composed of one type of cell, but the vascular system is much more than an association of identical channels. Instead, vascular strands consist of a number of tissues – xylem, phloem and generally also an embryonic cambium between them (Esau, 1977; Fahn, 1982). Each of these tissues, furthermore, consists of a number of cell types, and these cells occur in predictable relations and locations. The very same local sources of auxin induce the differentiation not only of vessels, the specialized systems that are most readily seen, but also of the sieve tubes of the phloem and the cambium between them. Phloem may require less auxin for its initial induction (Aloni, 1987a, 1987b), but this alone does not account for the quantitative relations of the two adjoining tissues. The complexity of the vascular system must mean that canalization is not the entire story – but not that the canalization hypothesis is necessarily wrong. Little can be said at present about the controls of the patterning of the various components within the vascular system, but some evidence can be found for the following four possibilities (Fig. 6.7).

(a) Specialization for different signals

Perhaps the most obvious possibility is that the complexity of the vascular system reflects a specialization of the different cells for different inductive signals (Fig. 6.8A; Sachs, 1972a, 1981a). The close association of the components of the entire system would result from an initial common specialization, and it is here that auxin could play a major role. Later, the different tissues would each transport their own specific signal, this transport being associated with specialized differentiation.

Available evidence concerning inductive signals other than auxin is very limited. There have been claims that sugars, together with auxin, promote the formation of phloem rather than xylem (Wetmore and Rier, 1963; Northcote, 1969; Waren Wilson, 1978). Although sugar transport is a major role of the mature phloem, induction by sugar is not necessarily expected: leaves cause phloem differentiation long before they become exporters of sugars. The evidence, furthermore is suspect; the reported studies have not been accompanied by the careful anatomical work required to recognize phloem. More recent work (Aloni, 1980) indicates that it is the formation of callose that is promoted by the added sugars, not the differentiation of the cells in which this callose is found. Since callose is a polysacharide, the enhancement of its formation by sugars in the substrate would hardly be surprising.

Gibberellins are the one group of hormones in addition to auxins that have repeated, qualitative effects on cell differentiation in the vascular

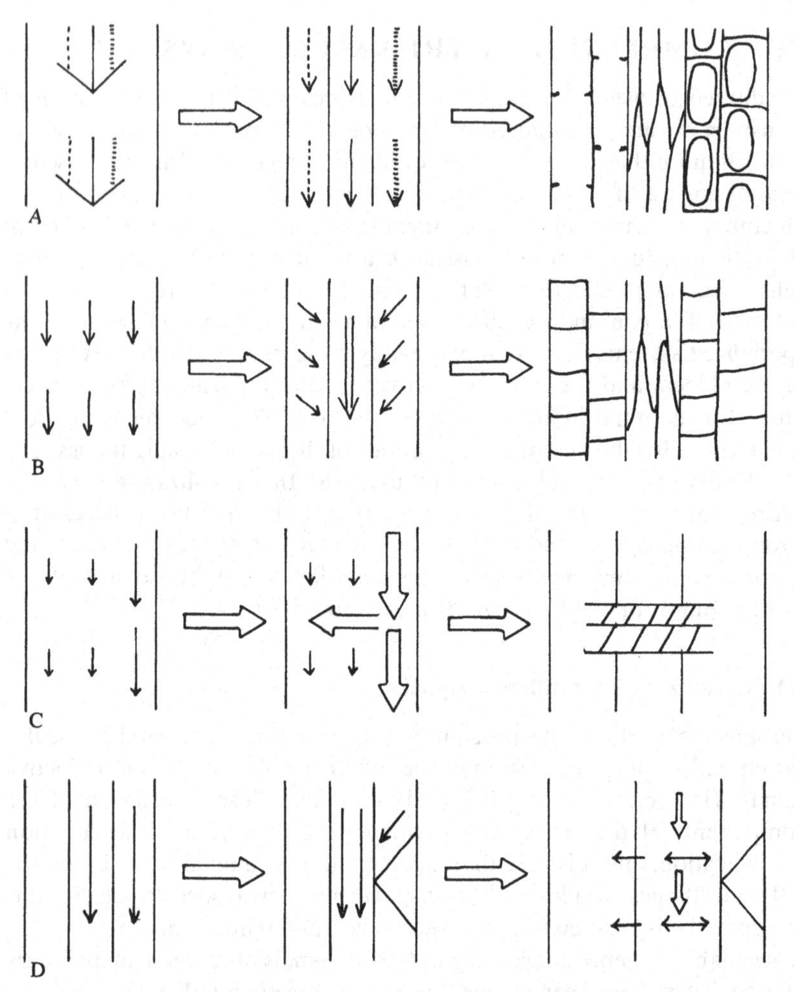

Figure 6.7. Schematic representation of 4 possible mechanisms which could determine the differentiation of more than one cell type within a vascular strand. These possibilities are not mutually exclusive and there is evidence that all of them have a role in vascular differentiation. A. The initial specialization of the strand cells is for the joint transport of a complex signal. During later differentiation neighboring regions specialize for the transport of the various component signals. B. The initial specialization involves many cells. Continued specialization limits signal flow to the central cells, leaving their neighbors 'partially induced' and thus specialized relative to both uninduced and fully induced cells. C. The differentiation of the components of a strand may be complementary: they may depend on one another through radial interactions. D. The specialization of one tissue as a transport channel may divert developmental signals away from its neighbors. This would be expressed when the transporting tissue is damaged and signals are diverted towards its neighbors. Such relations could control the relative development of the various component tissues. The horizontal arrows indicate growth in girth.

system. Gibberellins promote cambial activity and there have been indications that they specifically enhance phloem differentiation (Wareing et al., 1964). The available evidence, however, does not include careful microscopy and does not apply to many species of plants (Aloni, 1988). A more general, well substantiated case can be made for the promotion of fiber differentiation by gibberellins, especially when they are applied with auxin (Hess and Sachs, 1972; Aloni, 1976). New fibers can be induced in some tissues, especially when the apical meristems are damaged or are cut and re-grafted (Sachs, 1972a). These fibers are induced by signals of young leaves and the inductive effects follow the shoot-to-root polarity of the tissues (Aloni, 1978, 1979, 1987b). Taking these facts together, it may be suggested that during the early stages of fiber differentiation the cells are specialized transporters of gibberellins, a transport that does not continue when the cells mature. At present this would be only a reasonable working hypothesis within the framework of the ideas of signal canalization.

(b) 'Partial induction' determines parenchyma differentiation

The canalization hypothesis outlined above suggests that the cells which differentiate as vascular channels are selected from many potential possibilities, all of which undergo the initial stages of differentiation (Fig. 6.7B). This means that the processes of vascular induction should include cells that are 'partially induced', and these could have distinct mature phenotypes, quite different from those of the associated transporting channels (Sachs, 1972a, 1981a). Since the vascular channels are embedded in parenchyma, this parenchyma is an obvious candidate for the role of such 'partially differentiated' cells. The parenchyma occurs both as an essential component of the vascular tissues themselves and of the cortical and pith tissues that surround them.

The suggestion that the parenchyma represents axial tissue that underwent partial induction to become vascular tissues is supported by three lines of evidence. (a) As mentioned above the removal of very young leaf primordia, at the time they are first formed, prevents the formation of both the vascular strands and the parenchyma that would have surrounded them (Fig. 6.4 D–F; Wardlaw, 1946; Sachs, 1972a). (b) Removal of leaf primordia somewhat later causes the future vascular strands to mature as elongated, parenchymatous cells. 'Added' or grafted leaf primordia do not join rapidly enough for their effects on young cells to be studied, but it is possible to continue the induction of parenchyma cells with auxin. Young parenchyma, close to apices, is most readily induced by auxin to become either xylem or phloem (Sachs, 1981a). The inducible parenchyma cells are capable of transporting fluorescent dyes along their axis (Fisher, 1988). Finally, (c) the parenchyma of the plant axis is often more elongated, more polarized, the closer it is to the vascular tissues. This would be in agreement with the prediction that it is this parenchyma that was induced for the longest time before its flow of auxin was diverted away by the continued differentiation of its neighbors. It is also the cells closest to the transporting channels that differentiate most readily to form new vascular contacts around wounds. Taken together, these facts certainly point to much if not all of the parenchyma being 'partially induced' vascular tissues – but they by no means rule out the possibility that there are also unknown inductive signals special to parenchyma cells.

(c) Radial interactions

Another possibility, for which there is structural evidence, is that the differentiating cells of the vascular system interact and induce one another across the radius of the plant axis. This would mean, for example, that signals from the phloem would be necessary, in addition to the polar flow of auxin, for the differentiation of nearby xylem (Fig. 6.7C). The converse, a dependence of phloem on xylem differentiation, is less likely since small amounts of phloem commonly form in the absence of any xylem (Aloni and Sachs, 1973). A polarity across the radius of stems is indicated by anatomical structure, including the presence of rays, as well as by regeneration and grafting experiments (Fig. 6.8): tissues join along the radius of the plant only when their phloem–xylem axes correspond (Warren Wilson and Warren Wilson, 1981, 1984).

Two types of radial interactions have been suggested. The first is that differentiation depends on radial gradients of critical substances. The formation of the phloem, at least, could be associated with the formation and maintenance of this gradient and it would thus influence the differentiation of neighboring tissues. Warren Wilson and Warren Wilson (1961, 1984) suggested a detailed hypothesis of this type, based on the assumption that the precise concentrations of auxin and sugars are critical for phloem, cambium and xylem differentiation. This hypothesis has also been subjected to quantitative analysis by computer modelling (Warren Wilson, 1978).

A second possibility is that differentiation depends on the continued flow of signals across the radius of the plant. The ray system that connects the differentiating phloem, through the cambium, with the differentiating xylem would be a visible expression of such induction (Carmi et al., 1972; Sachs, 1981a). This ray system is missing only in plants in which there is very limited cambial activity; in all other plants rays are maintained and their size and number increases as the circumference of the plant increases. These rays are formed in response to active induction, rather than being 'left over' when the longitudinal system is induced: when plants are wounded, the space available for shoot-to-root contacts is limited and yet the proportion of the rays increases (Fig. 6.8 C–E; Carmi et al., 1972).

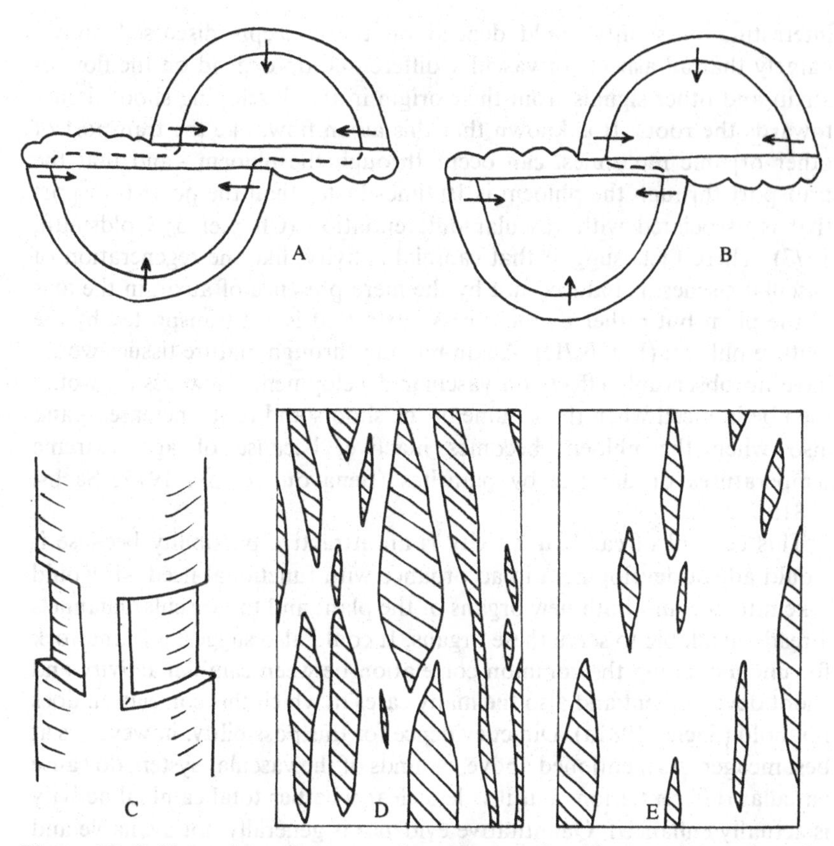

Figure 6.8. Evidence for radial interactions between the phloem and the xylem. A, B. Grafts of two stem halves lead to the regeneration of a continuous cambium. This regeneration always maintains the phloem–xylem orientation in neighboring regions, as in (A); contacts that would include a reversal of this orientation (B) are never formed, even when they would be the shorter alternative. C–E. In partial girdles around tree trunks the surface of the available cambium is limiting for further tree development. Such cambia develop rapidly, forming large volumes of xylem and phloem. Tangential sections of the wood show the proportion of the tissue occupied by the rays; this proportion is high in the rapidly developing bridges across girdles (D) as compared to untreated regions (E). The difference is large, as much as 100 percent. The formation of these rays is at the expense of the space available for transporting channels and it indicates that the rays have an essential developmental role – perhaps they are responses to a necessary movement of signals between the developing phloem and xylem.

(d) Signal diversion and the relations between the cambium and the mature tissues

Young leaves and other sources of auxin induce both the activity of the cambium and the differentiation of phloem and xylem. This raises the question how the different effects of the developing leaves are coordinated and how the size or capacity of the mature system is determined. An interesting possibility could depend on the concepts discussed above, namely that all aspects of vascular differentiation depend on the flow of auxin and other signals from their origin in the developing shoot tissues towards the roots. It is known that this auxin flow, like the transport of other organic molecules, can occur through the phloem – and that the transport through the phloem is 10 times faster than the polar transport that is associated with vascular differentiation (Chapter 5; Goldsmith, 1977). These facts suggest that cambial activity, like the regeneration of vascular tissues, is induced not by the mere presence of auxin in the axis of the plant but rather by the excess auxin that is not transported by the mature phloem (Fig. 6.7D). Auxin moving through mature tissues would have no observable effects on vascular development. New tissues would then be induced when the volume of the shoots and roots increases – and also when the phloem becomes inactive, because of age, extreme temperatures or damage by parasites (Benayoun et al., 1975; Sachs, 1981a).

This control of cambial activity is an attractive possibility because it would adjust development in accordance with functional need – it would take into account both new organs of the plant and the vascular channels already available to serve these organs. It could also suggest a framework for understanding the common correlation between cambial activity and shoot development and also the many cases in which this correlation does not hold (Sachs, 1981a). Direct evidence for this possibility, however, is at best meager. As mentioned above, wounds of the vascular system do cause vascular differentiation: but it is less clear whether total cambial activity is actually enhanced. Quantitative evidence is generally not available and wounds can be expected to have complex effects on differentiation because they reduce shoot growth. The limited quantitative evidence that is available (Benayoun et al., 1975) does show that wounds can increase total cambial activity even some distance away. This might also be an explanation for increases of cambial activity in response to damaging radiation (Chandorkar and Dengler, 1987).

Careful anatomical studies also point to the need for cambial activity that replaces entire vascular channels when only a limited location is wounded (Eschrich, 1954; Benayoun et al., 1975). Here, again, it appears that functional phloem, as long as it is present, reduces cambial activity. The replacement of entire phloem channels does not occur in all conditions (Schulz, 1986a,b): it is possible that young sieve tubes, which have not completed their differentiation at the time of wounding, can be reconstituted around wounds. This exception does not contradict the generalization that functional phloem acts to reduce cambial activity by diverting away inductive signals. This would be a competitive relation between two different tissues.

CONCLUSIONS CONCERNING VASCULAR PATTERNING

The central concept to emerge is an elaboration and continuation of the hypothesis suggested in the previous chapter: the specialization of one axis is an early and essential cellular process for the organized differentiation of plant tissues. The orientation of individual cells is expressed, induced and maintained by the flow of auxin, and presumably also by other, unknown, developmental signals. The influence of this flow is gradual, so that it becomes canalized to distinct files of cells as differentiation continues and the transport capacity of individual cell files increases. Auxin flow that can span the entire plant can thus serve in local interactions between neighboring cells: for the conformation of the axis and for the differentiation of cells of a file as well as in assuring complementary rather than identical differentiation of neighboring cells through which the flow does not continue.

Induction by long-distance hormone flow can be a basis for understanding the determination of the patterns of vascular strands. Differences between organs and between plants could be due to differences in the parameters of the formation of auxin and, especially, of the axial response of the cells to the flow of this auxin. But the pattern of the vascular system also reflects the growth of primordia: and early primordial stages have the greatest effects because strands, once they are induced, tend to divert auxin flow through themselves and to be maintained even when the entire shape changes dramatically.

The axial response of the plant cells could also be the basis for the complex cellular organization within the vascular system. At this level the evidence is specially meager. Only outlines of various possible mechanisms, which are complementary rather than mutually exclusive, are possible at present. It seems likely that cellular complexity could result from differentiation in response to flow for different lengths of time. It is also possible that the cells become specialized for the axial transport of various components of an inductive flow. There must be interactions between differentiating cells; the vascular rays may be an anatomical expression of such interactions. Finally, the very same signals that induce active differentiation may have no developmental effects when they are transported through mature tissues.

7

Cell lineages

THE DEVELOPMENTAL SIGNIFICANCE OF CELL LINEAGES

Plants are constructed from cells or divided into cells. Because of their thick walls, these cells are readily seen when tissues are magnified. Cells are units of inheritance: each one carries the entire genetic system of the organism and it is only in them that these systems are copied. Cells are also units of gene expression: within each one a type of differentiation prevails, while neighboring cells can be radically different from one another, even when they are the products of the division of one mother cell and are connected by numerous plasmodesmata (Chapter 9). The purpose here is to consider whether, or to what extent, cells and cellular events could also be units of the development of form.

From the time a cell is formed to the time it matures it can increase in volume by a factor of a thousand or more. Yet cell growth is always limited; continued development is dependent on the formation of new cells by the process of cell division. Any given tissue is thus the product of the divisions of an original, meristematic cell. Thus, a given mature cell must be the final product of a series of cell divisions, a *cell lineage* (Fig. 7.1). This lineage is shared to various degrees with other cells: some but not necessarily all neighboring cells are closely related ontogenetically, having originated from the same mother cell. Lineage is treated here as distinct from growth and from the relation between growth and cell division (Green, 1976). It is also considered separately from questions of developmental rates and the roles of promeristems and quiescent centers (Chapter 10).

Cell lineages are part of any full description of development. They could be indications of the ways in which form is determined and could be especially relevant to two related topics:

(a) A leaf, a branch or any other structure could arise from one or from a number of cells (Fig. 7.2). The presence of one mother cell would suggest, though it would not prove, that the determination of the

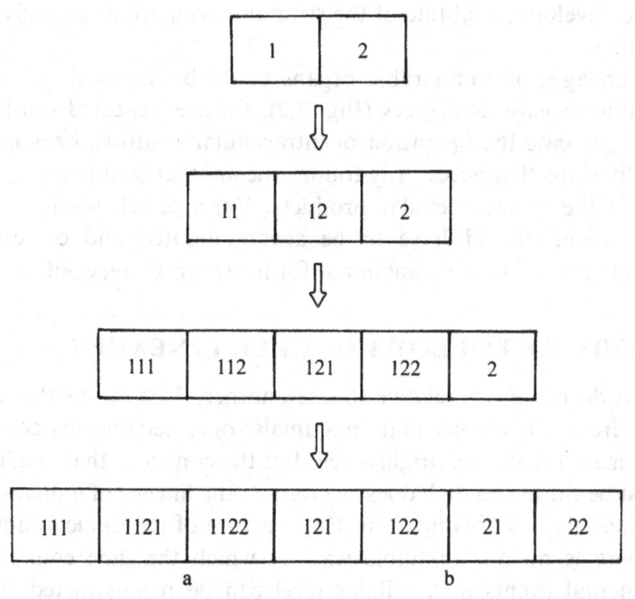

Figure 7.1. The concept of cell lineage. Cells are necessarily the products of a chain of cell divisions. In this schematic figure, the adjoining cells in location 'a' are the products of a recent division while the cells in location 'b' have been separated for a longer period. In the mature structure, there is often no simple way of distinguishing between these possibilities.

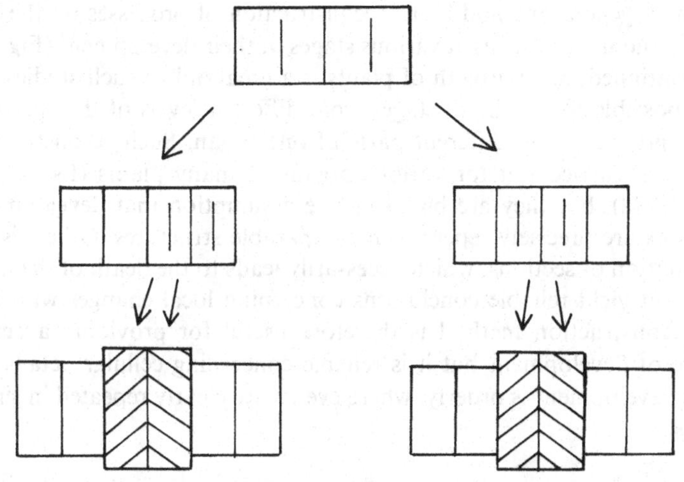

Figure 7.2. A variable cellular origin of comparable structures. A special, 2-celled structure arises on the left from one mother cell. This cell becomes central as a result of the inequality of the divisions on its two sides. On the right an identical structure arises as a product of the coordinate development of two neighboring cells. Both possibilities may occur within the very same organ and they may be indistinguishable in the mature state.

entire developmental fate of the structure occurred at an early, single cell stage.

(b) Cell lineages of comparable organs could be identical or could be variable to various degrees (Fig. 7.2). Precise, repeated cell lineages could indicate the operation of intracellular controls, ones in which cell divisions that necessarily follow one another lead to a determined fate of the various cellular products. Variable cell lineages, on the other hand, would have to be accommodated and corrected by cellular interactions if functional forms are to be reached.

METHODS OF FOLLOWING CELL LINEAGES

Plant cells do not move relative to one another. This means that lineage relations are much simpler than in animals: only neighboring cells could have originated from one original cell. But the converse, that neighboring cells must be closely related, does not follow; the lineage of individual cells is still generally not obvious from the structure of the tissue at any given time. There is no one infallible way by which the time course of the developmental events at a cellular level can be reconstructed. Various specific cases can be elucidated by the following methods, and their combined results do yield a general picture.

(a) Reconstruction of development from sections

The most general method is the reconstruction of processes on the basis of the structure of tissues at various stages of their development (Fig. 7.3). The continued, axial growth of plants is a great aid to such studies: it is often possible to obtain all stages from different levels of the very same plant and even from different parts of one organ. Such reconstructions have been carried out for various organs of many plants (Esau, 1977; Fahn, 1982), but they are based on the assumption that developmental processes are precisely repeated in comparable structures. Otherwise the examination of sections, which necessarily leads to the death of the tissue, could not yield reliable conclusions concerning local changes with time. The reconstruction method is therefore useful for providing a general picture of development but it is reliable concerning cellular details only where development is orderly, where events are clearly repeated in similar organs.

(b) Microscopy of living tissue

In special cases it is possible to follow the changes of living tissues without any damage or influence to the normal course of development. This has been done with small, transparent roots that could be kept alive and

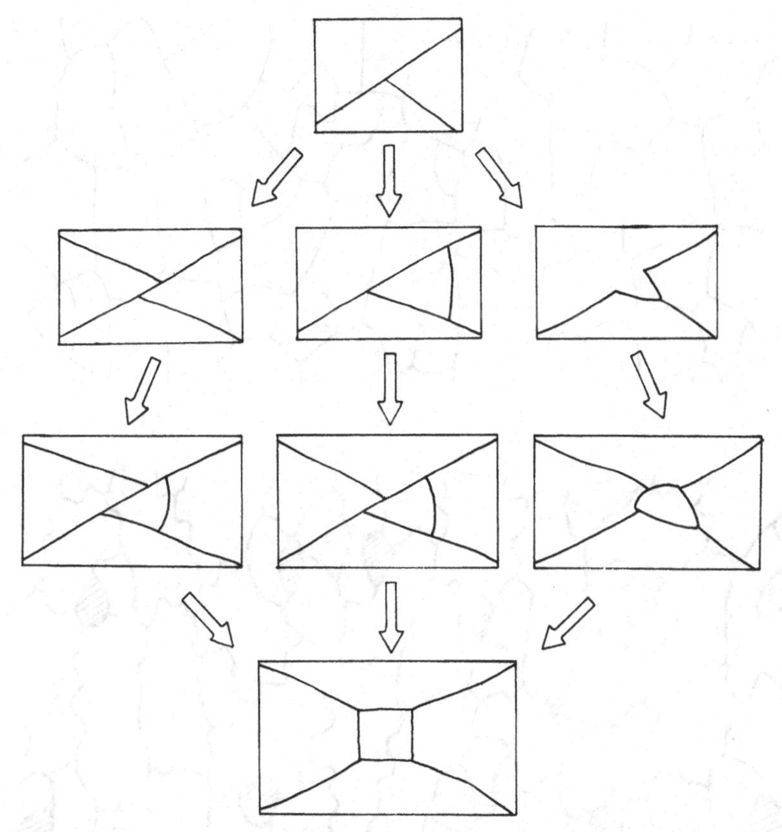

Figure 7.3. Reconstruction of a developmental sequence from stages in development of comparable structures. The 5-celled structure at the bottom of the figure could arise from the one at the top in various ways. The presence of varied developmental stages in young tissues suggests hypotheses such as the three shown. On the left is a sequence that requires an unequal cell division In the central sequence unequal growth of the various parts of the cells leads to a small, central cell. On the right the specialized, central region is formed by the intrusive growth of one cell between the wall of its two neighbors. Real data for a reconstruction of development appear in Figure 8.1.

growing while being observed under a microscope (Sinnott and Bloch, 1939; Hejnowicz, 1959). The development of surface, epidermal tissues can also be studied by epi-illumination microscopy, using light reflected from the surface rather than transmitted through other cell layers. The developing plant can be kept in normal conditions, and if tissue heating is avoided, there is no reason to assume that the actual observations influence development in any way. Such observations are laborious and the available information is not extensive (Ball, 1960; Sachs, 1978b, 1979; Green and Poethig, 1982; Venverloo et al., 1983). The cellular relations of

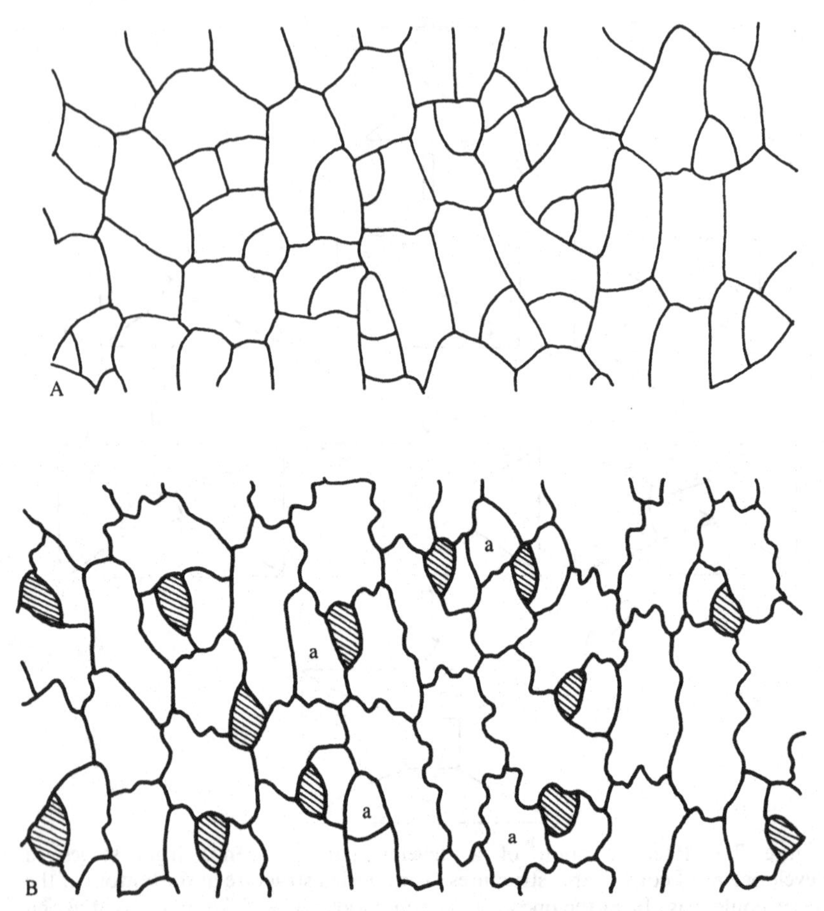

Figure 7.4. The development of a specific epidermal region. The surface of a developing stipule of a pea seedling was copied repeatedly using dental silicon (vinyl polysiloxane impression material). This treatment had no observable effects on the continued development of the stipule and the contours of the developing cells could be readily distinguished in the copies of the epidermal surface. In this way it was possible to follow changes in specific regions, identified by their unique cellular configurations. A. An early stage in which many unequal divisions can be distinguished. B. A later developmental stage of the very same region; the cells covered with diagonal lines are on their way to dividing and differentiating as stomata. Cell marked by the letter 'a' were similar products of unequal divisions, yet they were in the process of maturing as normal epidermal cells. The changed fate of these cells shows that it is essential that the development of specific regions of living tissues be followed. (Based on Kagan and Sachs, in preparation.)

the surfaces of many plants can be conveniently copied as plastic impressions: the borders between the cells can be recognized because they are depressed relative to the cells themselves. Repeated impressions of silicone preparations that polymerize without damaging the tissues are

Figure 7.5. Leaves of chimera plants, consisting of cells that differ in their genetic constitution. Such leaves are common in garden plants. The inability to form chlorophyll marks the cellular products of given initial cells; the different shades are due to different numbers of layers that do and do not form chlorophyll. It is characteristic of these chimeras that the precise borders between the different cell types are variable, both at the level of the whole leaf and at the level of the individual cell layers. A. *Hedera helix*. B. Leaflet of *Scheflera venulosa*.

perhaps the most promising method of following surface development (Fig. 7.4; Williams and Green, 1988).

(c) The fate of genetically marked cells: chimeras

A third method of following cell lineages uses genetic markers that are passed from a given cell to all daughter cells and are never passed to mere neighbors. This is possible where the cells are not genetically uniform, i.e. the plant is a chimera (Fig. 7.5; Weiss, 1930; Neilson Jones, 1969; Tilney Bassett, 1963, 1986). Use is made of chance mutations, such as those influencing the formation of chlorophyll (Stewart and Dermen, 1975). Such chimeras are often used as garden plants, where they are maintained by vegetative reproduction. It is also possible to induce observable mutations by irradiation of appropriate genotypes, where chromosomal aberrations result in observable chimeras. Any desired stage can be irradiated and the developmental fate of individual cells can be judged from the pattern of their products, which appears when the tissue matures as patches of special cells (Stein and Steffenson, 1959; Poethig, 1987). Methods are not yet available for influencing a specific cell, so conclusions must be drawn from the results of the statistical analysis of many random events.

(b) Cell lineages reflected in wood structure

Xylem is a stable tissue whose pattern does not change after maturation: the cellular patterns of the xylem reflect the state of the cambium at the time the specific xylem regions were formed. Serial tangential sections through the xylem thus contain a complete, detailed record of the changes that had occurred in the cambium (Bannan, 1951; Evert, 1961; Hejnowicz, 1961, 1967; Philipson et al., 1971). This is a possibility, unique in biology, of following the precise fate of individual meristematic cells over periods of very many years. The limitation of the method is that it applies only to the cambium, a highly specialized meristem.

THE PREDICTABILITY OF CELL LINEAGES

A central question raised above was the degree to which development is orderly at the cellular level, or, put differently, whether the cell lineage of a given structure is always the same. Repeated cell lineages have often been taken for granted as a logical necessity. Available facts clearly contradict this assumption. The most obvious expression of the variability of cell lineage is found in common garden chimeras (Fig. 7.5; Stewart and Dermen, 1975): tissues inheriting an inability to form chlorophyll vary greatly in extent. A demonstration of this can be found in a comparison of leaves along any stem or even in a comparison of halves of the same leaf. Comparable variability is found in numerous plants regardless of which tissues or layers lack chlorophyll, so variability could not be due to coupling between aberrant development and the albino traits. The variable fate of the descendants of any given cell is also indicated by the complex outlines of the descendants of individual cells in which mutations were induced during leaf development (Poethig, 1987).

The variability of cell lineage is also confirmed by the problems that arise when development is reconstructed from sections of tissues that could be expected to be comparable (Esau, 1977). Unpredictable details of cell lineages are also seen when the fate of individual living cells is followed: either directly, as in the formation of stomata in a developing epidermis (Fig. 7.4; Sachs, 1978b), or in the derivatives of the cambial initials, seen in the xylem (Bannan, 1951; Hejnowicz, 1961, 1967).

There are exceptions in which divisions are predictable or orderly. These are found in ferns, not in seed plants. The best studied case is the roots of *Azolla*, where a definite cell lineage can be established for the entire determinate organ (Gunning et al., 1978; Gunning, 1982). These exceptions to the rule that development is variable at a cellular level only mean that orderly development is possible; in no way do they contradict the conclusive evidence that in seed plants precise determination of the fate of developing cells is not common. The hypothesis that development must be orderly at the cellular level, which has been often accepted as an obvious, necessary assumption, is clearly wrong. The possible functions of variable details of cell lineages (Klekowski and Kazarinova-Fukshansky, 1984a,b; Klekowski, 1988; Sachs, 1988b) will be considered in the final chapter.

THE CONCEPT OF INITIAL CELLS

Initials of growing apices

The continued growth of plants occurs at the apices and this growth is correlated with cell division. The size of the apices remains more or less constant: most cells mature and cease to play a role as sources of additional cells. In the long run only the products of one or a limited number of cells can be expected to populate the apices. The cells that continue to be the source of meristematic cells, the ones whose divisions always yield at least one product that stays in the meristematic region, are known as initial cells (Fig. 7.7; Newman, 1956; Sussex and Steeves, 1967; Steeves and Sussex, 1972; Mignotte et al., 1987), stem cells or founder cells (Barlow, 1978). It should be noted that the existence of these cells is a logical necessity, not an hypothesis or a factual observation. This raises questions of how many initial cells there are in various meristems and whether they have any special traits related to their unique function.

The concept of initial cells suggests that they should be looked for microscopically, and indeed such cells can be found in ferns and other pteridophytes (Fig. 7.6A): they have unusual shapes and sizes, are situated at the very tops of shoots and roots and they divide in different planes in a predictable, repeated pattern (Bierhorst, 1977; Gunning, 1982; Gifford, 1983). When fern apices are damaged, they continue to develop only when apical cells are regenerated by changes in other, neighboring cells. These traits are the ones expected for initial cells.

The situation in seed plants is different: no specialized initial cells can be found, neither by light nor by electron microscopy (Clowes, 1961; Steeves and Sussex, 1972). Yet the number of initial cells in shoot apices of seed plants can actually be counted (Satina et al., 1940; Tilney Bassett, 1986). The method of counting is based on the development of chimeras, in which cells differ by genetic traits that are passed on to all their progeny but not to neighboring cells. It follows from the definition of initial cells that if there were but one initial there could be no permanent chimeras: eventually the entire tissue, since it has but one source, would have a uniform genetic constitution. Stable chimeras with two cell types thus show that there are at least two initial cells, chimeras with three cell types prove at least three initials, etc: the larger the number of the component cell types in chimeras, the larger the number of initial cells required for their stable maintenance during continued development.

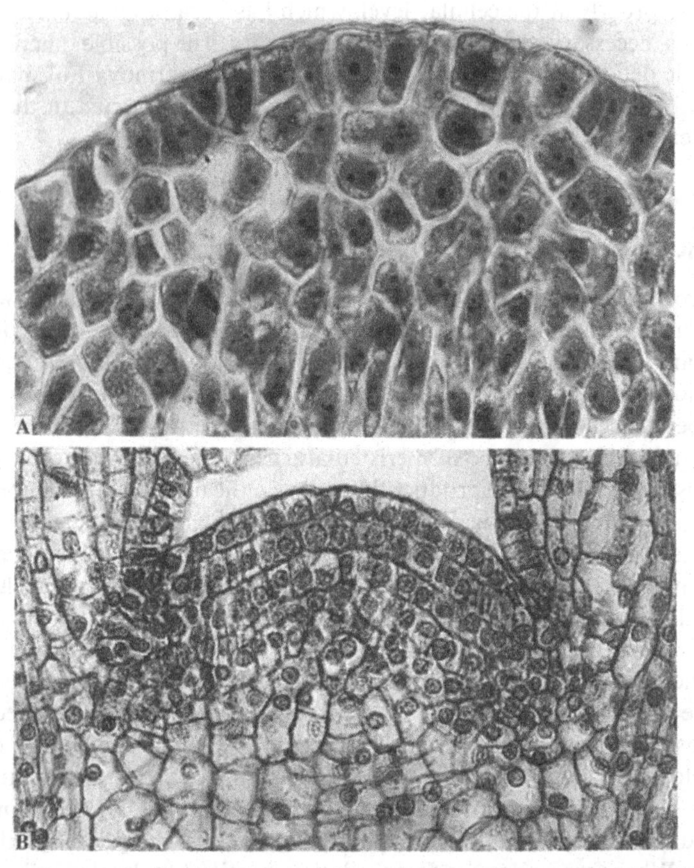

Figure 7.6. Longitudinal sections of the very tips, the promeristems, of shoot apices. Cellular configurations suggest the ontogenetic relations between cells. A. An apex of a *Lycopodium*. The cells divide both in parallel and at right angles to the surface of the apex. This means that cells at the very tip can be the ultimate origin of all the cells of the shoot. Cells with this special function, initial cells, may be recognized in the center of the very top of the apex. X 400. B. An apex of *Coleus*. The top cells are arranged in layers, indicating divisions only at right angles to the surface. This means that the cells of one layer do not contribute to another; there could therefore be more than one initial cell contributing to the long-term origin of the shoot. X 26.

Permanent chimeras have been obtained in various ways: by inducing random polyploidy with colchicine (Satina et al., 1940), by regenerating shoot apices along graft regions between tissues of different species (Weiss, 1930; Neilson Jones, 1969; Tilney Bassett, 1986) and by chance events that influence chlorophyll synthesis (Stewart and Dermen, 1975; Beardsell and Considine, 1987). Chimeras with three different genetic components, but no more, have been found repeatedly in various species.

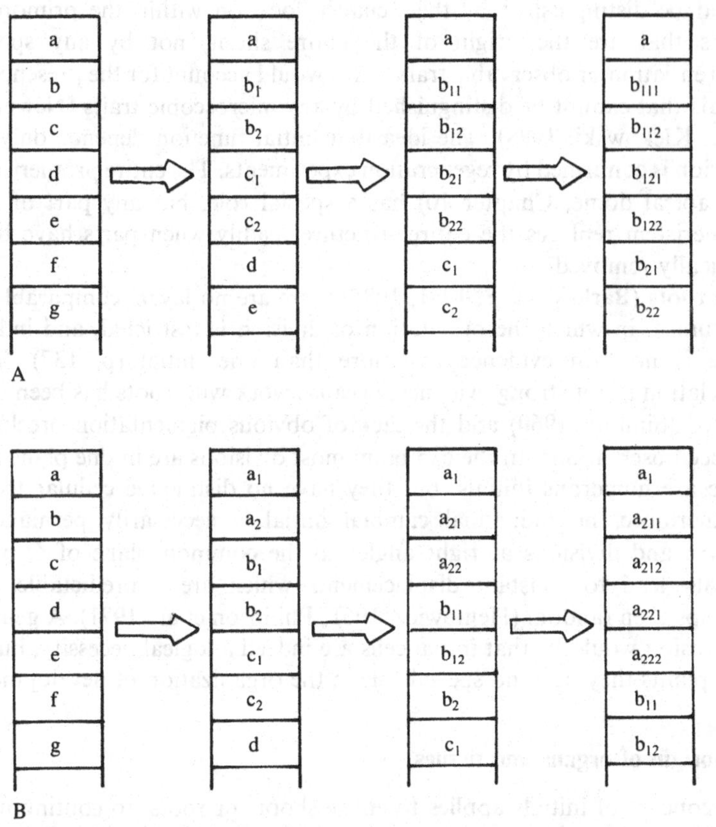

Figure 7.7. The concept of initial cells. Divisions occur only in the top 8 cells, the apex, of a uni-dimensional 'organism'. A. The top cell does not divide. The second cell, 'b', could serve as an initial. B. The top cell divides, though infrequently. The consequence of these divisions is that only the top cells can be a long-term initial of the developing tissues.

Such chimeras can be maintained for long or unlimited periods. It may be concluded that seed plants have shoot apices with at least three initial cells. Since more complex chimeras have not been found, it is unlikely that apices with larger numbers of initials are common.

The patterns of cell divisions in apices of seed plants are not regular and do not define any unique locations for initial cells. Yet divisions do follow patterns that could be the basis for there being more than one initial cell. The division of two or more outermost layers of the apical dome are consistently at right angles to the surface (Fig. 7.6B; Chapter 10). These oriented regions form the 'tunica' that covers the dome (Esau, 1977; Fahn, 1982), and they are presumably the reason that the cells of the outermost layers do not displace all the cell within the dome. Thus initials would be distinguished by their chance location within the primordial layers that are the origin of the entire shoot, not by any special differentiation or observable traits. This would account for the presence of initials that cannot be distinguished by any microscopic traits (Newman, 1956; Klekowski, 1988). The idea that initial function depends only on location is confirmed by regeneration experiments. The entire promeristem (the apical dome, Chapter 10) has a special role, but any part of this promeristem replaces the entire structure readily when parts have been surgically removed.

In roots (Barlow, 1978, 1981, 1984) there are no layers comparable to the tunica, in which the orientation of division is restricted, and indeed there is no firm evidence for more than one initial (p. 137). This correlation is not strong evidence, because work with roots has been very limited (Sinnott, 1960) and the lack of obvious pigmentation precludes chance observations. In the cambium most divisions are in one plane and there are numerous initials, but they have no distinctive cellular traits. Furthermore, no individual cambial initial is necessarily permanent: growth and divisions at right angles to the common plane of cambial activity lead to constant displacements which are unpredictable and perhaps even random (Hejnowicz, 1961; Philipson et al., 1971). A general conclusion would be that initial cells are indeed a logical necessity, but in seed plants they have no special role in the organization of development.

The origin of organs and tissues

The concept of initials applies to entire shoots or roots, to continuously growing systems, but the same idea could also be useful concerning determinate organs, such as leaves. Again, clear initials of entire leaves can be found in ferns and such initials are absent in seed plants. Microscopic observations of various stages in the formation of leaves of seed plants demonstrate the participation of a large number of cells, cells which are not related to one another by any unique cell lineage (Lyndon, 1979, 1983; Poethig and Sussex, 1984a; Cunninghame and Lyndon, 1986). The same conclusion is reached on the basis of chimeras induced in unstable genetic systems by radiation at suitably early stages of leaf formation (Dulieu, 1968; Poethig, 1987). These radiation-induced changes are random and rare, so they must occur in single cells. Thus the maximal size of regions affected by any one mutation is the size of the region produced by any one initial cell. In seed plants such induced mutations are only expressed in an entire leaf if they are also found in other, surrounding tissues. Mutations special to a leaf are always expressed in relatively small regions, confirming that as many as 100 cells contribute to the tissues of any given leaf (Poethig and Sussex, 1985b). Direct microscopy of roots developing from epidermal tissues suggest that they, too, can form from more than one cell (Venverloo, 1976).

There have been suggestions that reproductive organs are formed by specialized cells, set aside early in embryonic development. This idea is presumably derived from animal systems, in which the separation of the germ plasm is an important, though not a general, principle. Thus, the concept of 'meristèmne d'attente' (Buvat, 1955; Steeves and Sussex, 1972) suggested that cells in the center of shoot apical meristems do not divide or change until the apex becomes reproductive. Available facts are not in accordance with this suggestion. Though the central cells do not divide rapidly (Chapter 10), there is good evidence that they do divide (Steeves and Sussex, 1972), and even rare divisions are sufficient to contribute to vegetative tissues and to displace and change any cell within the central region itself (Fig. 7.7). There is thus no reason to assume any unique, predetermined group of cells with a special function in apical development.

Some results of studies of induced mutations in maize and cotton embryos would appear to contradict this conclusion concerning the absence of early determination of the cells that form reproductive structures. It has been found that the fate of the products of apical cells in these embryos could be predicted, within wide limits (Coe and Neuffer, 1978; Johri and Coe, 1983; Christianson, 1986). Yet it is possible to understand these results without assuming any predetermination that is not found in other plants. The predictable cell fate does show that in maize the cellular course of development is relatively repeatable or orderly, a conclusion not surprising for a monocotyledon with a small, determinate number of leaves (Chapter 12). Where cell divisions follow a predictable path, a cell could form predictable structures even if it were not predetermined or differentiated in any way. A pre-determination could be shown by an inability to form other structures when development is perturbed by wounds that lead to regeneration, but results of experimental perturbations are not available. Furthermore, recent work shows that at least in some conditions the variability of cell fate is true for maize as it is for other plants (McDaniel and Poethig, 1988).

Not only organs but also specific tissues and cell types could be the products of divisions of predetermined meristematic cells. An old concept of this type is that 'Histogens', or meristematic regions, are responsible for the formation of specific tissues (Sinnott, 1960). Histogens were described for various apices, and their terminology still appears in current discussions. It is true that because the outermost layer of shoot apices generally divides only at right angles to the surface this layer has the exclusive function of forming the epidermis (Bruck and Walker, 1985). It is also true that central regions of any shoot are the products of apical cells somewhat closer to the center of the apex. Beyond these facts it is hard to find generalizations that are valid even for the same plant under varied conditions, let alone different species and perturbed apices. This is true for roots as well as for shoots (Barlow, 1982, 1984). Furthermore, even where cell lineage is precisely repeated, as in fern roots, there is no simple relation between lineage and the patterns of differentiated cells (Gunning et al., 1978; Gunning, 1982): the same mature cells are formed by different lineage sequences in different parts of the organ. The concept of tissue differentiation being pre-determined in apical cells is thus unsupported by any facts of general validity.

CONCLUSION CONCERNING THE ROLE OF CELL LINEAGES

In seed plants there is a large variability in the cellular details of any given mature structure. This variability extends to all developmental stages. Cellular variability may have important advantages for long-lived meristems, since it allows cell selection which avoids the somatic accumulation of unfavourable mutations (Klekowski and Kazarinova-Fukshansky, 1984a,b; Klekowski, 1988). Variability of development at a cellular level is also characteristic of the development of many animals, though not that of the relatively simple nematodes (Sulston et al., 1983).

The facts are not consistent with an early determination of developmental fate of cells. Nor are the facts consistent with a rigid developmental program. Thus, at least some aspects of patterning are at a supracellular level. Although cells are the units of division and of gene expression, it does not follow that they are necessarily the units of all aspects of development (Cusset, 1986). At least some important controls are global – they modify development but do not specify individual stages at a cellular level. Such controls could, for example, stop the growth of a region once it has reached a pre-determined size, regardless of how this size was reached. These possibilities are considered in a wider context in the final chapter.

8

Stomata as an example of meristemoid development

THE CONCEPT OF MERISTEMOIDS

Stomata, hairs, secretory glands, sclereids, etc. are all examples of a broad class of structures that is difficult to define. It is their formation rather than their mature structure that is easily characterized: it involves a departure from the axial organization of a developing meristem (Chapters 5 and 10) through the appearance of a new, local polarity. The specialized cells formed by these departures from overall polarity cannot be simply related to the 'canalized induction' of vascular tissues and the polarization of the plant axis (Chapter 6). Because emphasis traditionally has been on mature tissues rather than on their development, there is no general term to cover all these structures; 'idioblasts' (Esau, 1977; Fahn, 1982) comes close, though it is generally limited to single cells and is also used in relation to specialized cells that run along the plant axis, such as laticifers.

A common early stage of cellular development that departs from axial or polar organization is an unequal cell division (Fig. 8.1; De Bary, 1877; Bünning, 1953, 1957). It is always the smaller product of such divisions that forms a specialized structure. The microscopic appearance of the small cells is distinctive: they have a dense cytoplasm and small vacuoles, characteristic of apical meristematic cells. The small cells are meristematic also in their ability to divide one or more times. These centers of specialized differentiation were named *meristemoids* by Bünning (1953, 1965); he defined them as isolated locations of maintained embryonic potency, and thus of continued divisions, in an otherwise maturing tissue. Yet it is very common for unequal divisions to occur in tissues where most if not all other cells continue to divide, forming cells of equal size and often of equal fate (Sachs, 1978b, 1979). The idea that meristemoids maintain an 'embryonic totipotency' is also hard to defend: meristemoids do not form all tissues but rather highly specialized structures which do not change under known circumstances. Furthermore, some of the structures that Bünning referred to as meristemoids, such as ray initials and differentiating xylem vessels, were isolated from neighboring events only when viewed in cross-sections and were not necessarily meristematic.

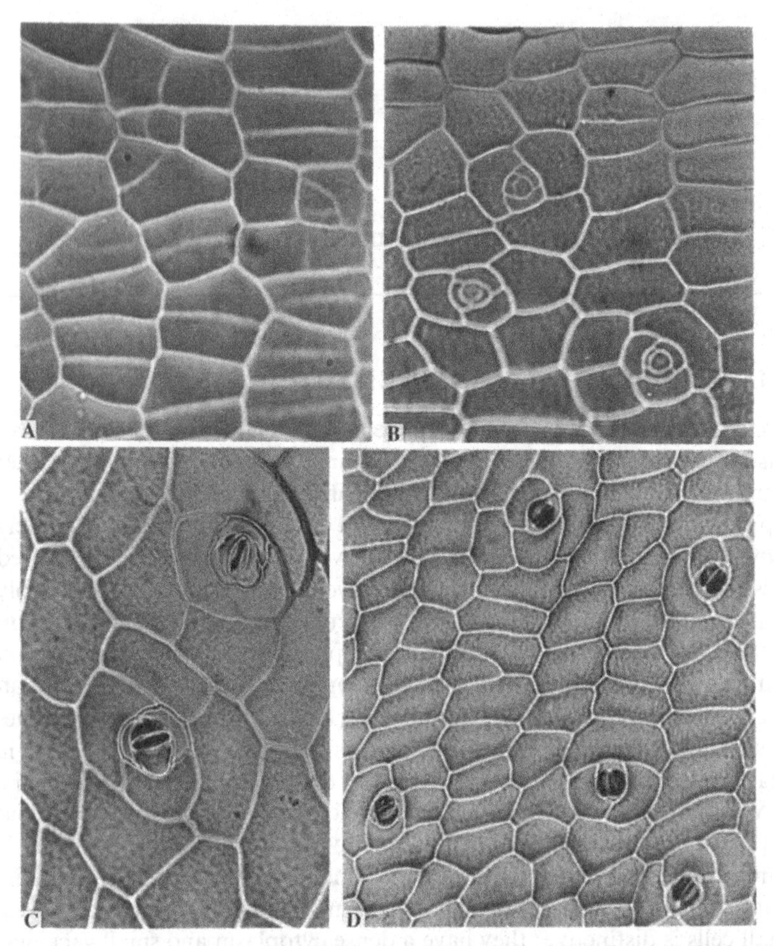

Figure 8.1. Stages in the development of stomata in *Graptopetalum*. Nail polish copies from a young and a mature leaf. It is easy to suggest a developmental sequence on the basis of these stages. It is also possible to see that the development of individual stomata does not follow any strict, repeatable course. A. Early stage, including future stomata that have undergone between one and three unequal divisions. X 350. B. Advanced stage, but before any stoma has undergone the final division that forms the two guard cells X 200. C. Two mature stomata that could have formed by divisions in two neighbouring cells. X 100 D. A more general picture, showing that the stomata are spaced: i.e., that they do not occur very close to one another. Otherwise the location of the stomata is quite variable. X 50.

Yet Bünning coined a useful term for a real class of structures, even if his definition needs modification. The term meristemoid will therefore be used here to refer to a cell that undergoes special differentiation, becoming distinct and departing from the axial organization of the tissues that surround it on all sides.

Meristemoids that appear microscopically similar to one another can develop into a wide variety of cell types (Bünning, 1953; Barlow, 1984). The fate of these meristemoids is a function of their location (Sinnott and Bloch, 1946) and of the timing of their development relative to the maturation of the surrounding tissue. The stress here will not be on the fate of meristemoids, that is, their differentiation as varied mature structures. This is a topic about which little can be said. Instead, the stress will be on the spatial relations, the pattern, of their cellular products. For reasons of brevity and clarity the main subject will be the meristemoids that form stomata (Fig. 8.2). These have been favored objects of research: they form in convenient locations on the plant surface, they are very common and their functional role is essential for most vascular plants. Here, again, the topic will not be the classification of types and the differences between species (De Bary, 1877; Rasmussen, 1981; Baranova, 1987), but rather the limited knowledge available about the determination of stomata patterns.

RELATIONS BETWEEN THE CELLS OF A FUNCTIONAL STOMA

As seen with a microscope, stomata consist of two guard cells and the pore between them (Fig. 8.2; Esau, 1977; Fahn, 1982). However, the functional system regulating pore size must include additional cells (Raschke, 1975). These neighboring cells may or may not appear different from the rest of the epidermis; where they do differ they were named subsidiary cells long before their function was known. Thus, a first question of cell patterning is related to the development of an individual stoma: how the spatial relation between the guard and subsidiary cells is established. This question is but an example of a very general one, since many of the structures formed by meristemoids include various cells with complementary functions and recognizable spatial relations. Adjoining cells which undergo complementary differentiation could develop in two ways, indicating two classes of developmental controls.

(a) The various cells may all be the products of one original cell which divides unequally. This could mean that the determination of the fate of the cells occurs very early, even before the original cell divides. In such cases the entire functional structure should always arise from one mother cell. Though unequal divisions are not necessarily obvious, it is likely that at least some of them would be recognizable by light or electron microscopy.
(b) The complementary possibility is that the fate of the neighboring cells is determined only at the time they are formed or even later. This could occur if cells of one type induce the differentiation of the other.

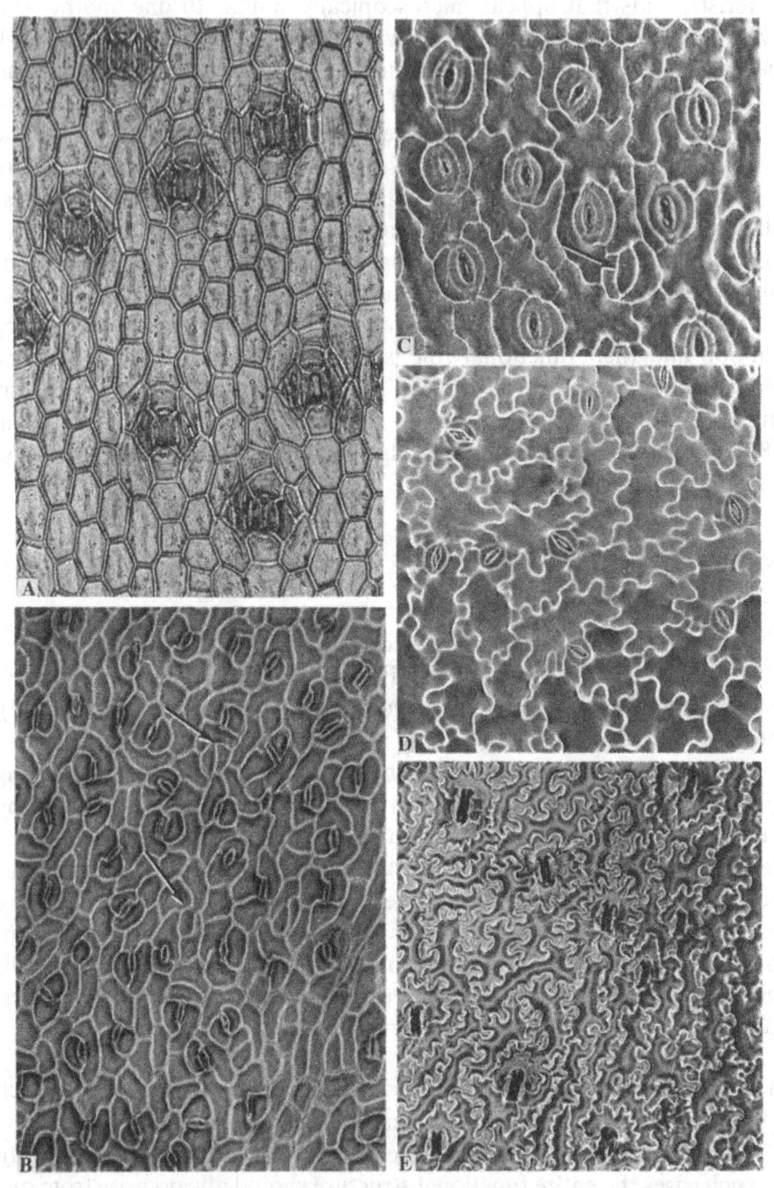

Figure 8.2. The epidermal structure of leaves of different species. Peeled cuticule of *Agava* in A and nail polish copies of the epidermis of *Lathyrus* in B, *Vinca* in C, *Impatiens* in D, and *Cyrtomium* in E. The stomata appear as pairs of small cells; their distribution within the matrix of other cells is not random. In some plants (A–C) there are also additional, subsidiary cells which adjoin each stoma. Though fully mature leaves were used, immature or aborted stomata (arrows) can be seen in *Lathyrus* and *Vinca*. X 90, 110, 300, 70 and 80, respectively.

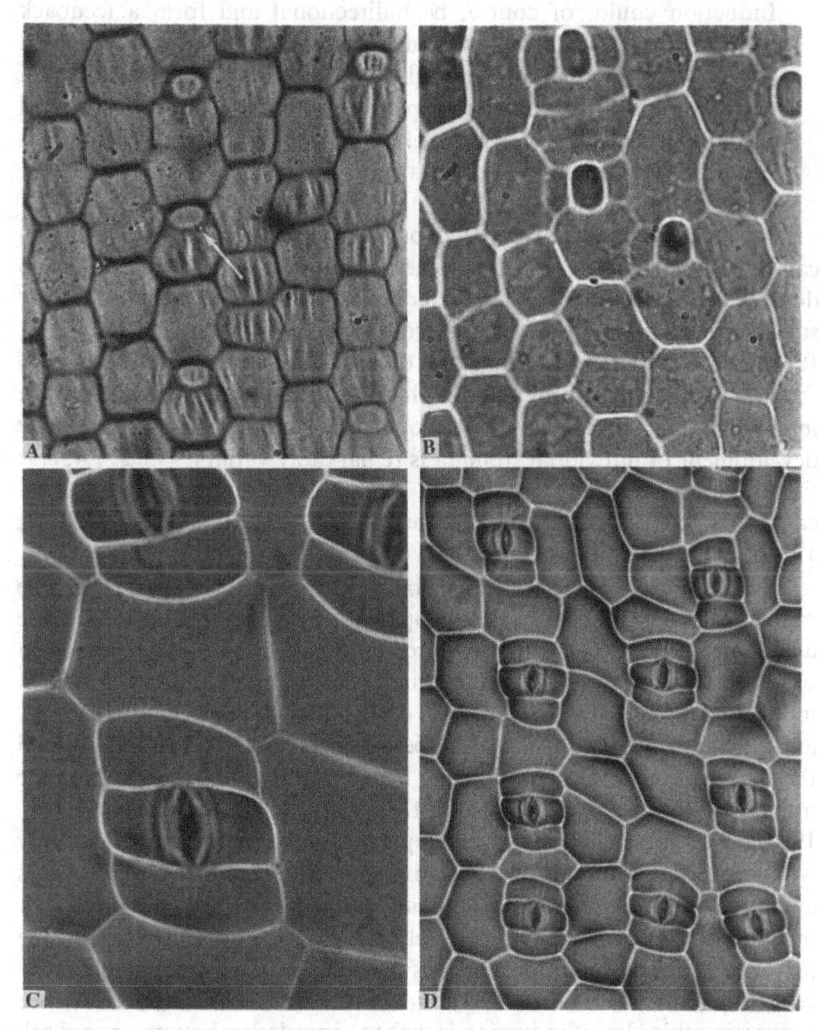

Figure 8.3. Nail polish copies of developmental stages of stomata of *Tradescantia*. Because of the way the leaf grows, the various stages were arranged along one axis at the base of young leaves. A. The first stage of stomata development is an unequal division, cutting off a small cell (arrow) towards the apex of the leaf. X 400. B. The developmental sequence suggests that the future stoma both induces and orients the unequal divisions in all the neighboring cells. Thus, as in many other monocotyledons, the subsidiary cells do not develop from the same cell as the stoma itself. X 400. C. Mature stomata that could have arisen from unequal divisions in neighboring cells. X 270. D. An epidermal field. Neighboring stomata are separated by the subsidiary cells and at least one additional epidermal cell. X 100.

Induction could, of course, be bidirectional and form a feedback control. This development would predict that the various cells need not arise from one mother cell. Furthermore, the developmental events should be amenable to experimental manipulations: if one cell is damaged during early development the differentiation of its neighbor would be modified in meaningful ways.

Major aspects of the development of stomata and comparable structures can be reconstructed from developmental stages (Chapter 7). There is no doubt that the entire functional complex, two guard cells and their subsidiary cells, is often formed from a single mother cell (Fig. 8.1; Rasmussen, 1981) and that unequal divisions are a characteristic part of their development. It thus appears that intracellular events could have a major role in meristemoid development. Furthermore, the cellular development of individual stomata is remarkably variable (Figs. 7.4, 8.1; Chapter 7). This may mean that intracellular adjustments to local conditions continue throughout the development of a stoma (Sachs, 1978b).

As might be expected, the general rule of the formation from one mother cell does not cover the development of stomata in all plants. There are stomata where the guard cells are accompanied by the differentiation of subsidiary cells of a separate origin. Examples of this are found in monocotyledons. In these plants it is easy to find stomata of different ages arranged in rows, the most mature being closest to the tip of the leaf and the earliest stages of development being next to, or even within, the intercalary meristem at the base of the leaf (Fig. 8.3; Stebbins and Jain, 1960). The course of stomatal development can be reconstructed not only with fair certainty but also with relative ease. And here the subsidiary cells arise not from the same meristemoids as the guard cells but rather by unequal divisions of neighboring cells (Fig. 8.3). This suggests that the presence of the stomata mother cells induces the division of neighboring cells (Stebbins and Jain, 1960; Stebbins and Shah, 1960). Evidence supporting this idea of induction is that the correlation between guard cell development and the divisions forming subsidiary cells holds even when the original meristemoids are in abnormal locations (Stebbins and Shah, 1960).

It is hardly surprising that the formation of complex structures from meristemoids involves various complementary processes. It is perhaps less expected that the evidence, as far as it goes, points to a restricted role of local inductive effects between cells. Furthermore, even where local inductive effects do occur the special traits of the induced cells could be due to their being young epidermal cells, to their competence to react (Chapter 1), and not to specific information conveyed by the signals of the developing guard cells. Thus inductive effects need not lead to a complex series of developmental events and there is no proof that they act as more than unspecific triggers. But, at least in monocotyledons, they do orient and localize essential events of cellular differentiation.

SPACING PATTERNS OF STOMATA AND THEIR MEASUREMENT

A second group of questions concerns not the relations between neighboring cells but those between neighboring stomata; these questions relate to the pattern of stomata distribution in the 2-dimensional matrix of the epidermis (Fig. 8.2). Stomata, like other products of meristemoid development, generally occur individually, and the first impression could be that they are randomly scattered. Bünning (1953, 1965; Bünning and Sagromsky 1948) pointed out that closer inspection shows that this distribution could not be completely random: though there are stomata in direct contact with one another, they are much less frequent than would be intuitively expected for a random distribution (Fig. 8.2). This means that the distribution of most stomata is a prime example of a 'spacing pattern' (Wolpert, 1971), a distribution in which distances are larger than would be predicted for random events.

For the study of spacing it is necessary to have measures of the orderliness of the system – of the deviations from random distribution that have to be accounted for by spatial controls of development. A variety of measures of distribution have been used concerning other spacing patterns (Lacalli and Harrison, 1978). Some of these methods have been based on ecological studies – the problems of evaluating the distribution of organisms are comparable to evaluating the distribution of cellular structures, in spite of the large differences in size. Yet the transfer of methods is not always straightforward: cellular structures such as stomata can occupy a measurable fraction of the area in which they are distributed, while the common ecological methods are designed to deal with points of negligible dimensions (Sachs, 1984b). The use of ecological methods has confirmed that the distribution of stomata is not random and these methods provided a quantitative measure of the orderliness of stomata distribution (Korn, 1972; Sachs, 1978b). These are important conclusions, but they do not suffice to characterize the mechanisms responsible for the distribution of stomata (Sachs, 1984b).

A measure that goes beyond mere degree of order is obtained by plotting the frequency of finding stomata as a function of the distance from a central stoma, chosen at random (Fig. 8.4; Korn, 1972; Sachs, 1974, 1978b; Marx and Sachs, 1977). This method is statistical; it involves the superposition of many stomata fields upon one another. In terms

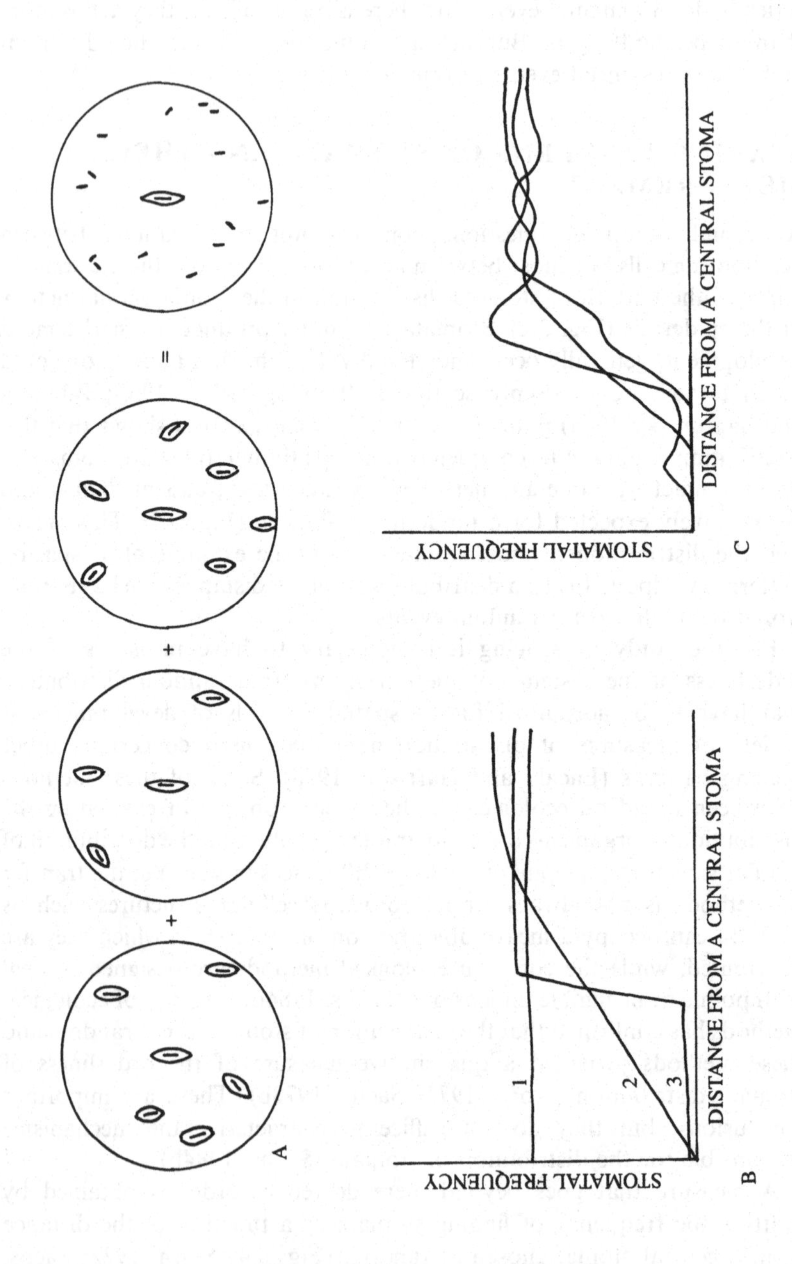
A
B
C
STOMATAL FREQUENCY
DISTANCE FROM A CENTRAL STOMA
1
2
3
STOMATAL FREQUENCY
DISTANCE FROM A CENTRAL STOMA

independent of any hypothesis, this method provides a graphic analysis of the relation of the development of a stoma and similar events in the surrounding tissues.

Measurements performed in many plants have yielded varied and sometimes complicated results (Sachs, unpublished). But some generalizations are still valid. The chance of finding a stoma next to another stoma is very low, much lower than would be expected for a random distribution (Fig. 8.4). This is what was pointed out by Bünning on an intuitive basis. The correlation between a stoma and neighboring events does not extend beyond the nearest neighboring stoma, a distance of one or a few cells in most plants. The change in the frequency of neighboring stomata is often abrupt: there is a 'stomata-free' region surrounding each stoma, and beyond this region the frequency of stomata is random (Korn, 1972; Sachs, 1974, 1978b; Marx and Sachs, 1977). It is only this last statement that is not readily apparent in preliminary measurements of many plants. The reason could very well be that the stomata are not formed together and different distances apply at different times, but this requires further study. Regardless of complications, the first aspect of stomata patterns that should be accounted for by any control of patterning is this 'stomata-free' region, in which additional stomata are very rare; this is the most orderly aspect of stomata distribution.

Figure 8.4. Method and results of analysing stomata patterns. The purpose is both to reveal the orderly component of stomata distribution and to measure the distances over which the presence of a stoma is correlated with developmental events in neighboring tissues. A. Three fields in which a stoma, chosen at random, was placed in the center. Such fields could be superimposed, as shown, for a statistical study of the relations between stomata and their neighbors. For reasons of clarity, the figure illustrates the superposition of only 3 fields; in actual measurements 50 or 100 fields could be used readily. B. On the basis of the distributions shown in (A) it was possible to plot the frequency of stomata per unit area as a function of the distance from any randomly chosen stoma. The figure shows 3 of the many possible curves: *1* would indicate no relation between the occurrence of a stoma and the frequency of other stomata in its neighborhood. This would mean that the distribution of stomata is random. Curve *2* would be obtained if a stoma has an inhibitory effect on the occurrence of additional stomata and the expression of this inhibition declined slowly with increasing distance. Curve *3* would indicate a stomata-free region surrounding each stoma. Beyond this region the occurrence of additional stomata is random. C. The forms of the curves found when the method was applied to real leaves; different curves were superimposed even though the actual distances from the central stoma varied greatly depending on the species and even environmental conditions. Random events are evident, but the general form indicates the presence of a stoma-free region, as in curve *3* in (B).

CONTROLS OF STOMATA SPACING

Bünning (1953, 1965) and Bünning and Sagromsky (1948) not only pointed out that stomata are spaced but also suggested a mechanism that could generate this spacing. Each developing stoma could inhibit the surrounding tissues from differentiating as additional stomata. If the inhibitory effects declined with distance – due to the dilution or metabolism of the inhibitory signal – then even initiation that is at first random could result in a mature pattern which is spaced. This hypothesis is reasonable, and mathematical modelling has shown that it could account very well for the observed phenomena (Gierer and Meinhardt, 1972; Korn and Fredrick, 1973; Meinhardt, 1982; Korn, 1981). It would predict other traits, such as the rare occurrence of neighboring stomata (Gopal and Shah, 1970) – where two neighboring stomata start developing at the same time – and the random components of stomata spacing. Mutual inhibition of developing stomata could thus account for some traits that have been used in arguments against its occurrence (Zimmermann et al., 1953). Yet the common attitude that mutual inhibition of developing stomata is necessarily true is not justified: it requires more direct proof and the consideration of other possibilities.

There is evidence that the controls of stomata patterning should be sought in the epidermis itself. They could not depend on the environment: it is not sufficiently fine-grained to determine the location of individual cells (Chapter 1). The environment can determine the overall density, as distinct from the pattern or relative location of individual stomata, even when density is measured relative to other epidermal cells and is thus independent of average cell size (Schoch et al., 1980). Nor could the pattern depend on events within the leaf, since no internal structure can be found to have a clear correspondence to stomata. The sub-stomatal chambers within leaves not only appear after stomata initiation but, importantly, do not have an invariable correspondence with stomata (Korn, 1972; Sachs, 1978b). This, of course, does not rule out other inductive effects of the mesophyll on the epidermis (Hake and Freeling, 1986). In the following discussion, different mechanisms of patterned differentiation within the epidermis will be taken up separately; they should be viewed as complementary rather than contradictory possibilities.

Packing mechanisms

Unequal cell divisions, the first observable stage of stomata formation, commonly occur in neighboring cells (Figs. 7.4, 8.1). As mentioned above, following development of the epidermis shows that guard cells do not form alone; the surrounding cells on one or even all sides are the products of the same original cell, the one which divided unequally (Fig. 8.1). It follows that intracellular processes of stomata development must be the basis of at least part of the stomata-free region, the region that is the important if not the unique aspect of patterning.

This calls for quantitative comparisons of the results obtained by the two methods: developmental studies, and measurements of mature stomata distributions. The clearest conclusions of such comparisons are found for plants of the family Crassulaceae, where a stoma is formed together with cells that surround it on all sides. In these plants the results of statistical measurements correspond very well with predictions of stomata-free regions made on the basis of developmental studies (Sachs, 1978b). One way of viewing this result is to consider development as forming multicellular units with a stoma at their center. Measurements then show that these units are 'packed' so that they can be in direct contact with one another. This does not mean that they are always packed with no separation of other epidermal cells; the measurements show only that the location of these epidermal cells is not patterned, and it might well depend on random processes.

In most other plants the stomata are formed together with neighbouring cells, but these do not surround the stoma on all sides. The size of neighboring cells accounts very well for the measured 'stomata-free' region (Sachs, 1974, 1978b; Marx and Sachs, 1977; Sachs and Benouaiche, 1978). Thus, there is good evidence that packing principles play an important role in most if not in all plants. But if packing of stomata complexes were the only factor, the stomata themselves could be in direct contact with each other on the sides not 'covered' by cells originating in the same mother cells as the stoma itself (Fig. 8.5). These contacts between stomata are rare, so additional controls must still be sought.

Evidence for cellular interactions

Patterning not accounted for by 'packing' of multicellular products of a single mother cell must involve interactions between cells. These interactions could be of different types: they could act over relatively large distances and polarize developmental events (Chapter 5) or they could involve local relations between neighbouring cells.

A polarity can be defined wherever stomata are associated with epidermal cells on only one or two sides of the complexes formed by a single meristemoid (Fig. 8.5). This raises the question of how this polarity is related to other polarities of the same tissues, especially the polarity of the vascular strands (Chapters 5, 6). The simplest, most consistent relations are found in narrow monocot leaves with a basal intercalary meristem (Sachs, 1974). Here all stomata complexes are arranged in rows, parallel with the vascular strands. Within each complex the guard cells are always on the side of the leaf apex (Bünning, 1957). This observation suggests that stomata development is influenced by the same long-

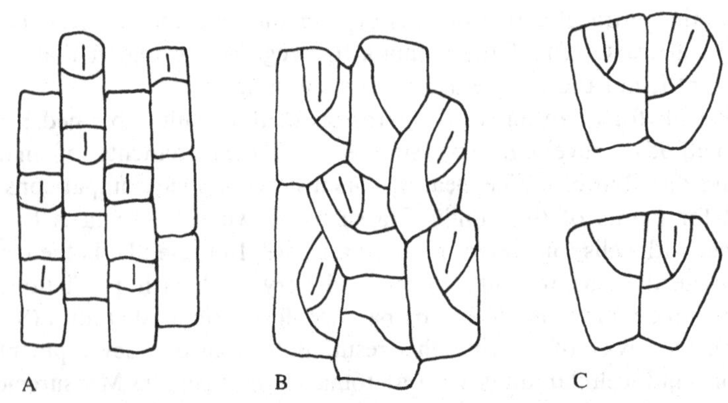

Figure 8.5. Stomata spacing resulting from polar stomata complexes. A. Each stoma forms a part of a complex, together with a basal cell. All stomata complexes have the same orientation, so along the axis of the leaf individual stomata are separated by at least one other cell. The cells in neighboring rows are staggered, again avoiding direct contacts between neighboring stomata. B. Stomata complexes which result from two unequal divisions. As long as these complexes have the same orientation they can result in a spacing pattern. There should be no exceptional stomata that are in close contact. C. Neighboring complexes as in B, except that their orientation would have led to the formation of neighboring stomata. In pea (*Pisum sativum*) leaves such events are followed by either oriented divisions, so that neighboring stomata are avoided (above), or by the development of a potential stoma as a normal epidermal cell (below).

distance, polar signals that influence the development of the entire organ (Chapter 5; Sachs, 1978b). This response determines the minimal distance between stomata: the sides of the stomata that have no associated epidermal cells do not face one another, and so immediate contact between the guard cells of neighboring stomata is prevented (Fig. 8.5A; Sachs, 1974).

Narrow monocot leaves are an extreme example; in most plants the polarity of stomatal complexes does not follow such simple rules. In these leaves all expressions of polarity have complicated, apparently inconsistent orientations: growth occurs in all directions and the vascular system is in the form of networks (Chapter 6). Yet the correspondence of long-distance polarity and stomata development appears to hold even for these leaves. Stomata have the same polarity as neighbouring vascular strands that differentiate at the same time (Goebel, 1922; Smith, 1935). The varied orientations of stomata could reflect their formation at different times: there are changes in the polarity even of a given tissue, as reflected in growth and in vascular differentiation (Chapter 6; Sachs, 1975a). This has important consequences for stomata patterns. Groups of stomata tend to form together and to have the same orientation (Sachs, 1978b, and unpublished results). This reduces the chances of contacts between stomata guard cells – but it does not prevent them where neighboring stomata do form at different times and have different polarities (Fig. 8.5C). Responses to long-distance, polar signals thus provide important though perhaps incomplete answers concerning stomata-free regions.

A possibility that requires study is that short-distance interactions change the fate of a stoma during development. These changes could be local and act to ensure that direct contacts are avoided. For evidence of such interactions it is not enough to reconstruct development from stages taken from various locations: the variability of events at this level precludes reliable results. It is therefore essential that specific regions be followed, and that this be done without damage or even disturbance to the developing tissues (Chapter 7). This is possible, though laborious, with epi-illumination microscopy (pp. 90–3). Copying the surface with silicone preparations (Williams and Green, 1988) may be a better alternative, only now being used for the study of stomata development. The available information is limited to very few epidermal regions of only two plant species (Sachs, 1978b, 1979). As far as this evidence goes, there are events that change development if it is about to lead to two stomata that would be in direct contact with one another (Figs. 7.4, 8.5C). The prevention of adjoining stomata occurs in two ways. A stoma that is very close to future neighbors may abort, the cells becoming regular members of the epidermis (Sachs, 1978b). The second possibility is for divisions to be so oriented that the future stomata become separated by a minimal cellular distance. Perhaps the control of orientation is also expressed by the gradual stabilization of microtubule systems during stomata development (Mineyuki et al., 1988).

These observations are at best indications that need verifying, but they do suggest the occurrence of local cellular interactions – interactions for which other, more compelling evidence was mentioned above in connection with the formation of subsidiary cells in monocotyledons (Fig. 8.3). Other indications of local interactions have been found in quantitative studies of the distribution of complexes of stomata and their neighboring cells relative to one another (Charlton, 1988). Related observations, in need of corroboration and extension to additional species, are that developing stomata may not mature when they are in a field with a sufficient density of mature stomata (Sachs and Benouaiche, 1978; Kagan and Sachs, unpublished).

ROLES OF CELL COMPETENCE IN STOMATA PATTERNING

The discussion above dealt only with stomata spacing and the stomata-free regions surrounding each stoma. Unusual distributions of stomata

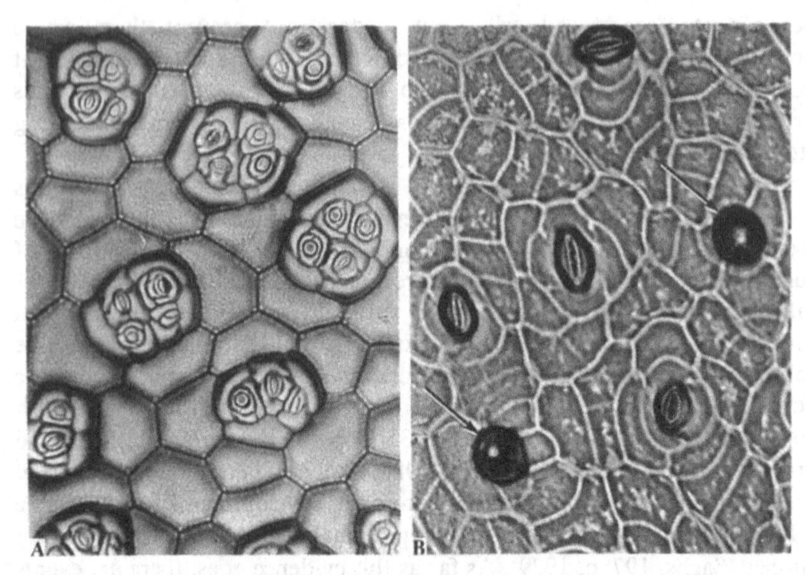

Figure 8.6. Examples of specialized stomata distributions. A. Stomata are arranged in groups in a *Begonia*. X 75. B. Hairs (round structures marked by arrows) are interspersed among the stomata in a *Peperomia*. The stomata and hairs form one pattern: they do not occur adjacently. X 200.

suggest another control of patterning. In some plants, especially in succulents, stomata may be arranged in groups rather than being spaced throughout the epidermis (Fig. 8.6A; Sagromsky, 1949). At least in some species, such as radish (*Raphanus sativus*) (Pant and Kidwai, 1967 and various begonias, the stomata in each group are added gradually from one mother cell. This may mean that the competence to form stomata is lost by most of the epidermis and is maintained in meristemoid cells; it is therefore only in these limited meristemoid regions that stomata continue to form. The result of this localized competence is groups of stomata (Sagromsky, 1949); within each group the stomata are spaced, and the pattern of cells suggests that this spacing is due to stomata forming simultaneously with their neighboring cells, as considered earlier in this chapter.

Another indication of a role of cell competence is found in leaves in which both hairs and stomata occur in the very same epidermis (Fig. 8.6B; Rasmussen, 1986a,b). Bünning (1953) suggested that since stomata do not form in the immediate vicinity of hairs, hairs must inhibit the formation of stomata in the neighbouring cells. Observations on the course of development of such plants (for example, *Pelargonium zonale*, Sachs, 1984b) suggest a related but different interpretation. Hairs form first, as centers of minute regions of the epidermis which mature early. When the stomata form, they are spaced throughout the epidermis except in the regions of the hairs which are relatively mature and do not take part in this later aspect of development. There is therefore no compelling need to assume that hairs actively inhibit stomata formation: it is possible that the cells in the regions of the hairs are not competent to respond to conditions which cause continued development, including stomata formation.

CONCLUSIONS AND THE PROBLEMS OF STOMATA DENSITY

The discussion above suggests the following scenario or working hypothesis. Stomata form as part of a complex structure, together with some of the neighboring cells. In most cases each of these complexes is an end product of processes that start in a single cell. The separation of the different functions within the complex is determined by intracellular processes, and these are followed by cell division and cell differentiation. In some plants – e.g., monocotyledons – the development of the stoma meristemoid can also have a polarizing or inductive influence, recruiting neighboring cells as part of the mature complex. In all plants, complexes that include a stoma can form from neighboring cells or meristemoids and can be in direct contact when mature: there is no evidence that they must interact with one another. The stomata complexes are polarized by the same long-distance interactions that influence other aspects of growth and differentiation; as a result of this polarization neighboring complexes are in contact on opposite sides and stomata rarely touch one another.

The surprising point, borne out by both the measurements mentioned above and by observations of development, is that the precise location of the cells that form meristemoids can be random, so that stomata initiation is not patterned and orderly distribution appears only during stomata development. There are indications of interactions between cells – but these, too, occur during rather than before the development of stomata. A characteristic of these interactions which could be important for future studies is that they influence not only cell differentiation but also the orientation developmental processes.

The capacity to form stomata, to become meristemoids, is limited to cells of a narrow 'competence window' (Wenzler and Meins, 1986) during the maturation of the primary meristem on the surface of the plant. There are no exceptions to this rule: stomata do not form from mature tissues nor from callus; conditions that induce a competence to form stomata, either *in vivo* or *in vitro*, are not known. The same is true for other meristemoids, including the ones that develop into secretory structures, whose formation in tissue culture could be of economic importance. Yet the variable density of stomata (Marx & Sachs, 1977) indicates that even in competent tissues not every cell forms a meristemoid:

Figure 8.7. The possibility of separate controls of stomata spacing and stomata density. In both fields the stomata are spaced as a result of a cell lineage forming a stoma together with the cells that surround it. Thus each square is competent to form a stoma, but stomata are not in direct contact even it formed by neighboring squares. Yet not all the potential positions of stomata are occupied. The frequency of occupied positions is higher on the lower field and this could be due to a higher frequency of random events leading to stomata formation, to a larger number of squares forming stomata, and not to a different patterning which would be a different relation of a stoma to its immediate neighbors. It is thus possible that density could be controlled independently of the spacing pattern.

the 'stomata-free' area that surrounds each stoma determines only an upper limit of stomata density, not the actual density found in leaves (Fig. 8.7). The proportion of cells that not only have but actually express the competence to form stomata could be a function of a frequency of cell change multiplied by the length of time that competence is maintained: and these factors, about which nothing is known, could be influenced by the conditions of growth and by the density of mature stomata within an epidermal field. Thus stomata density could vary depending on environmental conditions. Most of this variation is due to changes of the size and is not directly relevant to the topics considered here, but stomata density varies even when it is measured relative to the number of other epidermal cells (Schoch et al., 1980). The ratio of stomata to other epidermal cells is known as the stomata Index, and it is often more relevant to developmental questions than density related to epidermal area.

Are these conclusions valid only for stomata? An answer is hardly available, but to the extent that there is information it tends to support the applicability of the results to other systems. This is most obvious where the density of the specialized cells is high and the orderly aspects of the spacing patterns can be most readily discerned. The best example is root hairs (Bünning, 1957; Cutter and Feldman, 1973): they often originate as the smaller product of an unequal division and they are separated from one another only by one cell. On evolutionary grounds it is reasonable to assume that the same processes used in stomata development have diverged to form sclereids, secretory cells and a variety of other specialized structures (Sinnott and Bloch, 1946; Bünning, 1965; Barlow, 1984). There is hardly any evidence with which to test this possibility.

9

Expressions of cellular interactions

What do neighboring cells 'say' to one another? The answer is, of course, that we do not know. It is often accepted as self evident that local relations between cells must have a major role in organized development. It is assumed that signals passing between cells must be varied and specific, being responsible for the diversity of cell types. These assumptions require proof. The problem of obtaining evidence is that whatever the events, they occur over minute distances and are not readily amenable to experimental manipulations. Of course, the correlated differentiation of neighboring cells is itself evidence for interactions, but these correlations can often be understood without assumptions of local, specific interactions (Chapters 6–8, 10).

Additional evidence can be gleaned from various approaches. This chapter more than any other must therefore have a review aspect and deal with different topics, related to one another primarily by their implications for the general subject of cellular interactions. Thus, the microscopic examination of plant tissues reveals various structures that span two or more neighboring cells. Such continuity, the correspondence of special locations in different cells, could not be due to chance. The occurrence and development of the most common of these structures will therefore be considered. Another, related topic will be the developmental changes which occur when tissues are experimentally brought into new varied contacts, when originally separate cells are grafted together.

CYTOPLASMIC STRANDS AND WALL THICKENINGS

The structures spanning neighboring cells that are most readily seen are cell wall thickenings of regenerating tracheary elements (Fig. 9.1). The continuity of these strands across neighboring cells must be an expression of subcellular interactions between cells. These wall thickenings are preceded by cytoplasmic strands along the walls (Sinnott and Bloch, 1945; Hepler and Newcomb, 1963; Goosen-de-Roo, 1973). The cytoplasmic strands, furthermore, are often apparent even in cells which do not differentiate as vascular elements. Electron microscopic studies have

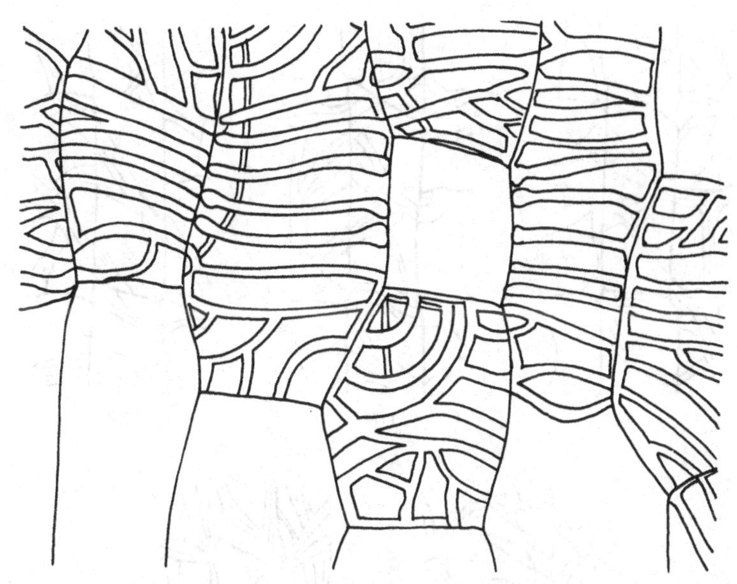

Figure 9.1. The continuity of wall structures in neighboring cells. Regenerative tracheary elements leading to a growing bud on a cut branch of *Impatiens sultanii* were viewed in thick sections using a projection microscope and continuous changes of the focal plane. The thick strands were secondary wall regions which contained lignin. In most cases these strands continued from one cell to another and this continuity means that there are local interactions between the neighboring cells. X 200.

shown that wall formation is preceded by microtubules (Hardham and McCully, 1982; Gunning & Hardham, 1982; Burgess, 1985), and microtubules are presumably the skeletal element of the cytoplasmic strands seen by light microscopy. The forces that pattern the microtubules and the cytoplasmic strands in these cells are not known, but they must span neighboring cells.

Cytoplasmic strands related to vascular differentiation have been found not only along the cell walls but also running through the vacuoles. These vacuolar strands are parallel to the future divisions and the long axis of vascular channels (Kirshner and Sachs, 1978). This relation is maintained even when vascular differentiation occurs around wounds, within various angles to the original axis of the cells. The available evidence would suggest that these strands are on early expression of events that span many cells, events that are presumably related to the canalization of auxin flow discussed in chapter 6.

Cytoplasmic strands can be observed *in situ* in epidermal cells in the transition region between meristematic and mature tissues. Here they do not precede any overt differentiation and their significance, if any, is unknown. These cytoplasmic strands are continuous from one cell to

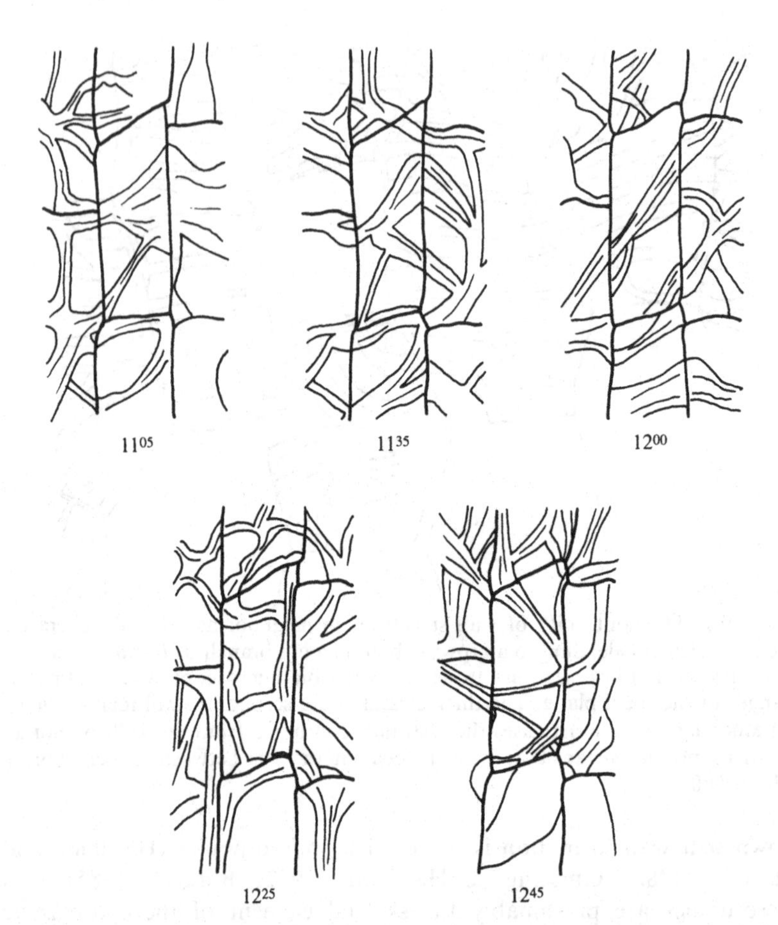

Figure 9.2. The continuity of cytoplasmic strands in neighboring cells and their temporal changes. The observations were of cells close to a living root tip, connected to an intact *Tradescantia* plant. The cells were in the early stages of elongation; unlike the cells of the actual tip they have large vacuoles and unlike cells further away from the tip the vacuoles are traversed by cytoplasmic strands. The drawings are of the same group of cells and they summarize the 3-dimensional structure, observed in many focal planes at the time indicated on the figure. The strands form a complex pattern which is generally continuous across cell walls and thus indicates local interactions between neighboring cells. This pattern changes with time – it is a manifestation of a dynamic, temporal organization. X 600.

another and it is only reasonable to assume that they express some interactions between the cells. But if this be true it is necessary to account for the lack of any consistent orientation in these strands: they can be followed from one cell to another, but no rules seem to apply to preferred orientations or the locations of thin and thick strands. It is possible to follow changes in these strands in living, small roots of *Tradescantia* (Fig.

9.2; Sachs, unpublished). Such observations show that the strands are in a state of flux, the entire pattern changing with a period of less than an hour. This suggests an hypothesis that deserves study: the random pattern at any given time may reflect a consistent pattern when it is integrated over a period of a few hours or perhaps days. This would mean a 'temporal organization' and different periods of cytoplasmic continuity could reflect the relative role of interactions in each direction. This could be important for tissue organization, but it is at best only a working hypothesis.

CONTACTS THROUGH CELL WALLS

Plant cells are separated by relatively thick cell walls. The contacts that can be seen through the walls are of three main types: plasmodesmata, primary pit fields and pits (Esau, 1977; Fahn, 1982). Plasmodesmata (Gunning and Robards, 1976) are thin cytoplasmic strands, 30–60 nm in diameter, bounded by cell membranes. The common occurrence of plasmodesmata means that the cytoplasm of plant tissues is part of a continuous 'symplasm' – that this cytoplasm is both continuous as well as bounded in small cellular compartments (Lüttge and Higgenbotham, 1979). The walls between cells are specially thin where there are groups of plasmodesmata and these regions are called primary pit fields. The third type of intercellular contacts, pits, are seen only in differentiated cells, ones with a secondary cells wall (Esau, 1977; Fahn, 1982). These secondary walls are interrupted so that neighboring cells are separated only by a thin primary wall and the pits are thus regions of facilitated passage of materials. Highly specialized regions of plasmodesmata occur in sieve tubes; very large pits from which even the primary cell wall has disappeared – actual holes – are found in the vessels of the xylem.

As mentioned above, the structure of these contacts through cell walls, their very continuity from one cell to another, must indicate interactions between neighboring cells. Furthermore, these contacts could presumably facilitate intercellular relations, and if their distribution is not random or their activity is not uniform, they could have a role in tissue patterning. Both the development and the function of contacts through cell walls are therefore central to any consideration of patterned cellular relations. But as determinants of cell fate, the stress must be on plasmodesmata only. Primary pit fields are associated with plasmodesmata and since by definition pits are found in cells with a secondary wall, which are cells whose developmental fate has been sealed, pits must be expressions rather than causes of differentiation.

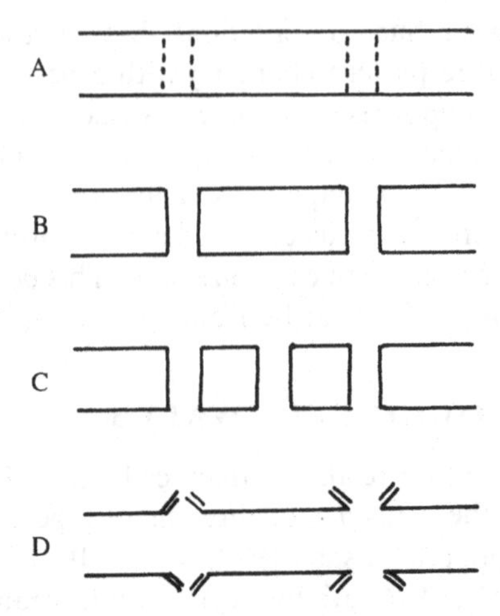

Figure 9.3. Four possible modifications of plasmodesmata which would influence intercellular communication. a. The plasmodesmata may be occluded during development. B. The density of the plasmodesmata may be reduced by the growth of the cell walls. C. New plasmodesmata may form through existing cell walls. D. The passage of substances through plasmodesmata could be modulated by temporary structural changes. The 'plasmodesma' on the left is closed while the one on the right is open.

Determinants of plasmodesmata density

Plasmodesmata are formed at the time of cell division: many cytoplasmic strands remain in the new cell walls, and these strands develop into plasmodesmata. As the meristematic cells grow, the density of the plasmodesmata is reduced. Counts of plasmodesmata in undisturbed tissues have shown a negative correlation between growth and plasmodesmata density (Juniper and Barlow, 1969; Gunning, 1978). This correlation results in different plasmodesmata densities in walls of different orientations: for example, since stems grow more in length than in width their cells will have more plasmodesmata in their transverse than in their longitudinal walls (Juniper, 1977; Gunning, 1978).

These facts suggest a minor role for plasmodesmata as controls of development – their occurrence and density would be a passive response to oriented growth, not a cause of developmental changes (Fig. 9.3B). But the early development of plasmodesmata during cell division is one major process, not the sole determinant of plasmodesmata and their activity. There is unequivocal evidence, from different sources, that plasmodesmata can form through existing walls. Thus new plasmodesmata have been shown to form in various pathological conditions (Burgess, 1972, 1985), but these special cases are not required to prove the point. The vascular cambium in the trunks of trees is a meristem that can remain active for thousands of years – and during this entire time the relations between its cells change constantly, new contacts being made as the result of intrusive growth (Fig. 7.3). Yet plasmodesmata connections between the cells are maintained in the cambium – these plasmodesmata can be seen and their development as sieve regions is essential for phloem function and thus for tree survival (Sachs, 1986). Further evidence for the formation of plasmodesmata through existing cells is their presence in graft chimeras (Burgess, 1972, 1985). These chimeral plants are formed in a region of a graft from cells of different genetic constitution, and they can develop, and reproduce by vegetative means, for unlimited periods (Chapter 7; Weiss, 1930; Neilson-Jones, 1969; Binding et al., 1987). Since these plants never had a single cell origin, their plasmodesmata must have formed across existing walls. Finally, there is no doubt that new plasmodesmata are formed across regular grafts. As in the case of the cambium, this is indicated by the presence of continuous vascular channels across grafts. Direct, elegant proof of plasmodesmata formation has come from grafts of plants of very different species (Kollmann and Glockmann, 1985). Here cells of different origin could be recognized even at the electron microscope level, the level required for the identification of new plasmodesmata. It is remarkable that in these grafts, which are only partially successful,, there are many 'half plasmodesma' – ones that are not 'met' by similar events in the neighboring cells (Kollmann et al., 1985).

There is thus no doubt that plasmodesmata can form between existing cells. There is also evidence that plasmodesmata can be occluded (Gunning, 1978): for example, stomata guard cells (Chapter 8) lack plasmodesmata when mature, and they develop from meristems where all cells are connected (Erwee et al., 1985; Palevitz and Hepler, 1985). What is less clear is whether plasmodesmata form and are occluded soon after guard cell division rather than during differentiation. In relation to this question it is relevant that a role of canalized, long distance signals in inducing plasmodesmata formation is indicated by the differentiation of the specialized contacts between the cells that are members of vascular channels. Additional controls of plasmodesmata activity are likely, but have hardly been proven. Perhaps grafts between tissues of different species (Kollmann and Glockmann, 1985) will turn out to be suitable for studies of the controls of plasmodesmata activity.

Can transport through plasmodesmata be modulated?

Turning to a functional aspect, plasmodesmata are the only channels through which molecules can pass form one cell to another without crossing two cell membranes. But this does not mean that the passage of substances through plasmodesmata is not restricted. Electron micrographs indicate that much of the channel through plasmodesmata is plugged by a dense cytoplasmic structure – the desmotubule. More direct evidence comes from studies of the movement through tissues of substances that do not pass through cell membranes (Goodwin, 1983; Erwee and Goodwin, 1983a,b), substances that must be injected if they are to enter living cells. Fluorescent materials of different molecular size have been used, so that movement to neighboring cells could be followed and compared *in vivo*. The results indicate that the largest molecules that pass through plasmodesmata have a size of 600–800 Daltons. It is remarkable that molecular sizes of the same order of magnitude pass through the gap junctions that connect animal cells (Overall et al., 1982).

A critical question concerning the role of plasmodesmata in controlling development is whether their transport properties can be modulated. Such modulation could occur within each plasmodesma if its structure changes, affecting its permeability (Fig. 9.3D). The structure of plasmodesmata includes a ring that could act as sphincter (Olesen, 1979), but concrete evidence for sphincter action is difficult to obtain because structure can only be observed by electron microscopy, after elaborate preparation. Following the movement of fluorescent materials has indicated differences between tissues (Erwee and Goodwin, 1985; Erwee et al., 1985), but these differences are not as large as could be expected if modulations of transport were important controls of differentiation (Santiago and Goodwin, 1988). Finally, the electrical coupling between cells should depend, at least partially, on the movement of ions through plasmodesmata. This coupling differs depending on the tissue and the orientation of the cells in question (Overall and Gunning, 1982), but clear evidence for the role of plasmodesmatal modifications is still lacking.

To sum up: there is no doubt that plasmodesmata are indications of cellular interactions and that they are likely to have a role in determining cell fate. What is less clear is the degree to which these effects can be specific and whether plasmodesma function can undergo fine modulations that precede rather than follow differentiation. At the present state of poor evidence, it might be a mistake to assume that plasmodesmata necessarily indicate highly specific interactions.

GRAFTS AS EVIDENCE OF CELLULAR INTERACTIONS

Grafts are a method of bringing together cells in any desired configuration: they are therefore straightforward experimental manipulations that could provide data concerning cellular relations. And grafts are common in the plant world. They have been an agricultural practice in the growth of fruit trees, at least in the Mediterranean region, for thousands of years (Vöchting, 1892). Grafts also occur naturally between branches of a tree that rub against each other in the wind. Grafts of roots are common, and may even be the rule in some species. The relations between higher plant parasites and their hosts can also be considered as a special form of natural grafts.

The central topic relevant here is evidence from graft success and failure concerning interactions between cells. These interactions can be roughly classified in two types: (a) ones that involve the passage of long distance signals through the region of tissue union; and (b) recognition events between cells in immediate contact with one another. This separation is at least partially artificial, but for the sake of clarity these types of interactions should still be considered separately.

Long distance effects on graft joining

There is good evidence that long distance interactions between organs (Chapter 2) are essential for joining of grafts (Fig. 9.4). Furthermore, these interactions are mediated by hormones. An anatomical expression of these interactions is the differentiation of vascular tissues (Chapters 5, 6; Lindsay et al., 1984; Yeoman, 1984). This role of vascular differentiation is not surprising, since vascular channels are the efficient bridge of the gap between the joined tissues. An indication of the importance of vascular differentiation is the difficulty of grafting monocotyledons: these plants do not have a cambium, so the formation of new vascular contacts that span the graft region would require changes in callus or other parenchyma cells over large distances (Chapter 6).

Direct evidence for the role of long-distance interactions in graft joining comes from experimental treatments in which various organs were removed, or removed and replaced by known hormones (Fig. 9.4). Young, developing leaves, morphologically above the graft region, greatly increase the proportion of grafts that join (Stoddard and McCully, 1980). As expected, the effect of the leaves can be replaced by an exogenous source of auxin (Parkinson and Yeoman, 1982). Agricultural practice shows that growing shoots which are not above the grafted region reduce or even prevent the success of grafts. Such shoots could be expected to compete with grafted tissues in the establishment of feedback relations with the rest of the plant (Chapter 3). Roots, on the other hand, are not essential for graft union in short-term experiments, provided the tissues are protected from desiccation. The importance of signals originating in developing shoot tissues is further indicated by grafts of two roots or two shoots (Fig. 9.4H,I). Two roots grafted together do not join, not even if they are connected to cotyledons and are supplied with all necessary resources. There is no problem joining two shoots in a similar situation, and vascular differentiation suggests that this joining is accompanied by local reversals of tissue polarity (Chapter 5).

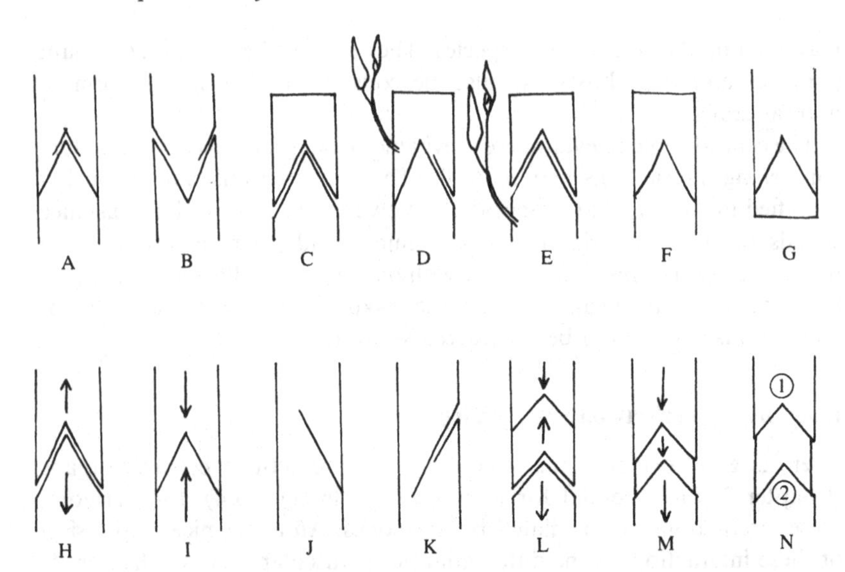

Figure 9.4. Evidence for the role of tissue polarity and long distance interactions in the joining of grafted tissues. The experiments were performed on pea (*Pisum sativum*) and other seedlings; consistent results were obtained when the grafted tissue were neither meristematic nor mature. Double lines indicate tissues which have not joined, while single lines in the region of the graft indicate successful tissue joining; arrows show the original polarity or root direction of the tissues, before they were cut and grafted. A, B. The grafted surfaces were diagonal at two different orientations relative to tissue polarity. Joining occurred primarily where polarity could be expected to concentrate endogenous auxin and cause this auxin to cross from one tissue to another. C. The shoot above the grafted tissues was removed. Unless the grafted tissues were very young this treatment prevented, or at least reduced, the joining of grafts. D. The growth of a lateral bud above the graft replaced the effect of removal of the shoot, at least on one side of the stem. E. Growing buds below the grafted region reduced rather than increased the chances of the graft joining. F. Auxin, applied as a lanolin preparation, replaced an intact shoot or a growing bud in promoting graft joining. G. The removal of the root system below the graft did not prevent tissue joining. H. Two root systems grafted together did not join. this was true even when the tissues that were in actual contact were part of the stem. I. Shoot systems grafted together joined readily, even though the two polar streams of auxin met at the graft region and at the time the tissues joined together the new plant had no roots. J, K. Diagonal cuts were made in stem and the tissue flaps were then tied back with parafilm. Downwards-pointing flaps, as in J, were most likely to join. L, M. Grafts of an intermediary tissue between the shoot and root systems. In L the polarity of this tissue was inverse to that of the other two graft members. As expected from treatment H, this inversion prevented the joining of the lower graft. In M the intermediary region was only a few millimeters thick. The polar effect of such thin intermediates was overcome: new shoot-to-root contacts were formed across both graft surfaces. N. Similar experiments on apple trees. The shoot and root systems, marked 1 and 2, were of different incompatible varieties, ones which did not join when grafted together. An intermediate region of a third variety could form two successful grafts and thus bridge the two incompatible varieties. In contrast to the previous grafts in this figure, this final result could only be understood as an expression of local interactions between the tissues that are in actual contact.

Indications of local recognition between cells

Positive evidence for long distance interactions certainly does not rule out recognition events between cells that are in direct contact with one another. The earliest stages of graft union are almost immediate, before any differentiation could occur: the wounded tissues stick together (Yeoman, 1984). However, this sticking occurs even between tissues that do not join in the long run, and it occurs even with foreign, non-living materials (Moore and Walker, 1981). It thus appears that the early sticking is due to temporary effects of substances released by the wounded cells. Critical events of cell recognition could only occur at a later stage, after the cells have been in contact for some time.

The joining of compatible tissues has been studied with both light and electron microscopes (Yeoman, 1984; Kollmann and Glockmann, 1985; Kollmann et al., 1985). Successful grafts have been compared with attempted grafts between tissues that are incompatible, tissues that do not join (Moore and Walker, 1981; Parkinson et al., 1987). These studies are valuable, but they have not yet resolved all the critical events between the cells that are originally separate and later become one tissue (Yeoman, 1984). Some local interactions must occur, since the debris of the cut cells disappears and, as mentioned above, plasmodesmata are formed across the graft region and the cytoplasmic membranes of the cells become continuous. There is also a building of cell walls, but wall fortification is a common response of plant cells to wounds (Moore, 1982).

Additional evidence concerning interactions between cells in grafts could come from restrictions of the specificity of the tissues that are able to join one another. This specificity should be considered both at the level of the species and at the level of the different tissues of an individual plant. As a general rule, the success of grafts increases the closer the taxonomic position of the plants whose tissues are being grafted. There are remarkable exceptions to this generalization. Some varieties of the same species, such as applies, cannot be grafted successfully. On the other hand and as mentioned above, there have been successful grafts of tissues of species that are unquestionably distant, such as *Helianthus annuus* and *Vicia faba* (Kollmann and Glockmann, 1985). Even the interpretation of the general rule in terms of local, specific cellular interactions is difficult (Moore and Walker, 1981). The failure of incompatible grafts could be due to disharmonies in the long-distance, hormonal interactions and not only the local incompatibility of cells. Yet there is evidence that requires local interactions: small bridging regions of one variety are used in agriculture so as to graft together shoots and roots of two other varieties that are not compatible (Fig. 9.4N; Yeoman, 1984). Perhaps the best systems for the study of short distance interactions will be grafts *in vitro*, using very small sections (Parkinson et al., 1987).

Grafts between different parts of the same plant generally succeed regardless of the original locations of the tissues and their relative orientations. In contrast to animal tissues (French et al., 1976), such grafts do not lead to the regeneration of missing tissues, those that 'should have been' between the graft members. Contrary to statements in excellent books on plant development (Bünning, 1953; Sinnott, 1960), the conformity of the original polarity in the grafted tissues is not required for tissue joining (Fig. 9.4I; Vöchting, 1892; Sachs, 1968). The source of the mistake is probably that polarity does influence the course of vascular differentiation and can prevent the formation of new contacts between the roots and shoots and thus lead to plant death; and yet local vascular whirlpools (Sachs and Cohen, 1982) join the grafted tissues. Inverted grafts of short regions, through which reorientation is possible (Chapter 5) are even used in agriculture, as a way of reducing vegetative growth and increasing fruit yield (Sax and Dickson, 1956). The grafting of tissues of different polarities, however, need not indicate a lack of specificity in relation to tissue joining but rather the facility with which local changes of this polarity can occur (Chapter 5).

Most types of plant tissues can be grafted and only three common exceptions need be mentioned. The first is mature pith or cortical tissues and is not surprising; inactive tissues, such as mature pith, are difficult or impossible to graft but the same cells are also unable to re-differentiate as new vascular channels (Sachs, 1981a). Epidermal surfaces do not join except in rare, specialized systems (Walker and Bruck, 1986). This could indicate a need for an exposed, wounded surface. Joining of unwounded meristematic tissues may occur during normal development of flowers (Walker, 1975), but it is an uncommon process, found only in few tissues in some plant families. The unexpected difficulty is in grafting promeristematic tissues, the ones that are considered to be 'undifferentiated' and are capable of unlimited, though slow, divisions (Chapter 10). There are a few reports of grafts of shoot tips in unusual locations (Gulline and Walker, 1957; Sachs, 1972a), but since tissue joining required long periods of time, it might have followed only after the cells differentiated or changed in some way. The difficulty of grafting promeristems could be related to their slow development and to their low level of participation in long distance interactions across the plant.

CONCLUSIONS

The main conclusion here must take a negative form: there is surprisingly little evidence for local, specific cellular interactions. Most of the available facts can be understood as local expressions of signals that act over long distances, signals for which there is other evidence (Chapters 2, 3). There are, of course, varied facts which do indicate the action of additional, local 'conversations' between cells. Yet it is remarkable that even known facts do not require highly specific interactions, ones that differ depending on the genotype, differentiation and orientation of the cells involved. This statement is suspect, because it concerns a lack of evidence. There is no doubt that plant cells are capable of elaborate recognition systems: these are expressed in relation to sexual reproduction (Cornish et al., 1988). Whether comparable specificity is required for patterned development must remain an open question, to be discussed further in the final chapter.

10

Apical meristems

Apical meristems are the site of the continued embryonic development that is an outstanding feature of plants. It is in them that organogenesis and the formation of new cellular patterns occur. It follows that an understanding of the development of form requires an understanding of the organization of meristems: it is necessary to seek structural subdivisions and controls that determine the orderly occurrence of developmental events.

Much of the organization of apical meristems must depend on processes internal to the meristems themselves (Ball, 1952; Buvat, 1955; Cutter, 1965; Steeves and Sussex, 1972). This is to be expected on the basis of the scale of the events, which would require detailed, fine-grained information if it were to be imposed by the environment or the rest of the plant (Chapter 1). The internal basis of organization is further suggested, if not proven, by the common development of new or adventitious meristems in unusual locations and even on unorganized callus. Furthermore, meristems develop normally when isolated by severe cuts (Wardlaw, 1952) and in culture (Murashige, 1974). These are conditions in which detailed organized information could hardly be supplied from outside the meristem itself.

The object of the present chapter is to compile a list of major organizing processes that occur in apical meristems. This will be at best a valid attempt, for not much is known. Yet some generalizations are possible on the basis of a combination of studies of structure, development, regeneration and responses to various treatments. These approaches should be most useful when information from many different plants and different conditions is used. Yet so as to simplify the discussion of a difficult subject, only the apical meristems of dominant, vegetative shoots and roots will be stressed.

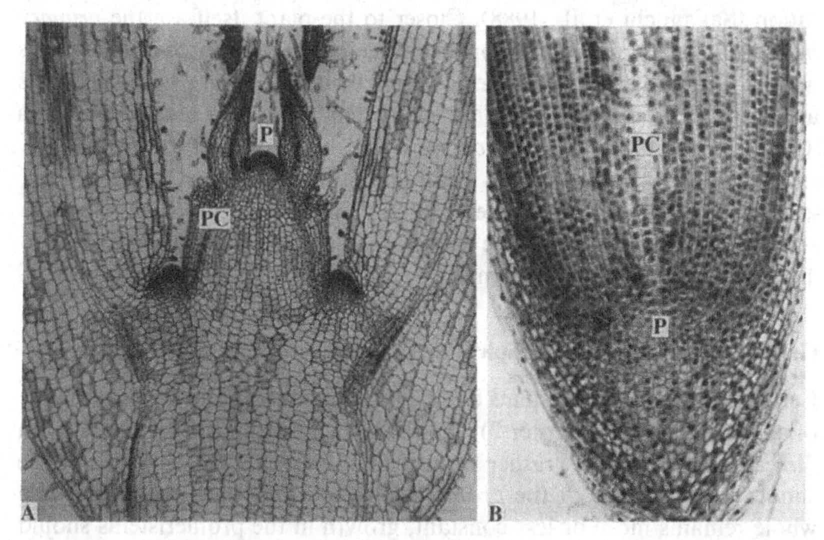

Figure 10.1. Longitudinal section through the embryonic apices of shoots and roots. 'P' marks the promeristems and 'PC' the procambium. A. Vegetative shoot apex of *Coleus blumei*. The rounded tip above the youngest leaves is the ultimate origin of all new tissues, even though it is not the site of the most rapid growth. The more basal the tissues the more mature they are. The uppermost, youngest leaf primordia are seen in both sides of the shoot tip. The next pair of leaves were not in the plane of the section and only some of their tissues appear, detached, above the apex. The third pair of leaves are large and only their bases can be seen, with axillary buds – the primordia of new shoot apices. The embryonic tip was protected by hairs and secretory structures seen in various orientations on the surface of the young leaves. Differentiation is seen primarily in the gradual changes in the cells as one proceeds downwards and in the presence of strands of the procambium – the future vascular tissues – connecting the leaves with the resting of the plant. X 9. B. A root apex of *Allium cepa*. Many features resemble a shoot apex; the major differences are the absence of leaves, the presence of a root cap covering the promeristematic tip and the central location of the procambium, whose cells are not necessarily elongated. X 8.

PROMERISTEMS AND PRIMARY MERISTEMS IN SHOOT APICES

Shoot apical meristems have characteristic cells: they are small, with thin walls, dense cytoplasm and no conspicuous vacuoles. In spite of these common characteristics cellular structure still suggests the division of apical meristems into two large regions (Fig. 10.1). Close to the tips are *promeristems* (Sussex and Steeves, 1967; Steeves and Sussex, 1972; Esau, 1977) in which the cells are not clearly differentiated from one another (Fig. 7.6) (possible sub-regions are mentioned below). The cells of the promeristems are not clearly polar and at least at the tip of the promeristematic dome even the microtubules have no consistent orientation (Sakaguchi et al., 1988). Closer to the plant itself are the *primary meristems*, in which cell divisions continue but at least three cellular types appear. The *procambium*, the region from which the future vascular tissues are formed, is characterized by elongated cells and specific enzymes (Gahan, 1981, 1988). The *protoderm* is the origin of the future epidermis and is readily recognized by its location, the shape of its cells and external secretions (Bruck and Walker, 1985). Finally, the *ground meristem* occupies the rest of the volume of the organ and it matures primarily into the various types of parenchyma.

Growth in various parts of apical meristems

Cell divisions in promeristems could be expected to be the ultimate source of all shoot tissues (Chapter 7). This follows not from their traits or from direct observations but rather from their location, at the very tips of the shoot: since the size of the promeristems and the apical meristems as a whole remains more or less constant, growth in the promeristems should displace the primary meristems towards the parts of the plant in which cells mature and cease to grow and divide. It is therefore critical to establish whether new cells are formed in the promeristems of shoots.

There is no doubt that promeristems are not regions of rapid growth, and the question whether they grow at all during their vegetative phase has been much debated (Steeves and Sussex, 1972). In shoot apices slow growth has been established by observations of living apices and by the presence of cells in various stages of division (Ball, 1960). The occurrence of divisions in shoot promeristems is also indicated by the small number of initial cells (Chapter 7): if promeristems did not grow there would be a large number of initials on the interface between the growing and quiescent tissues (Barlow, 1976). It follows that though not many cells are formed in promeristems, they are the ultimate source of new tissues, at least as long as new organs or other structures are formed. In contrast to promeristems, both cell division and increase in cell size in primary meristems can be very rapid.

The evidence of regeneration

Separate, complementary functions of promeristems and primary meristems are also suggested by their regenerative capacities. Regeneration of removed parts is relevant to the role of correlative signals in development and to the degree to which developmental differences have become determined: this is because the replacement of missing parts must mean that they had been the source of signals that prevented their duplication in the intact organs (Chapter 2) and that the state of the tissue

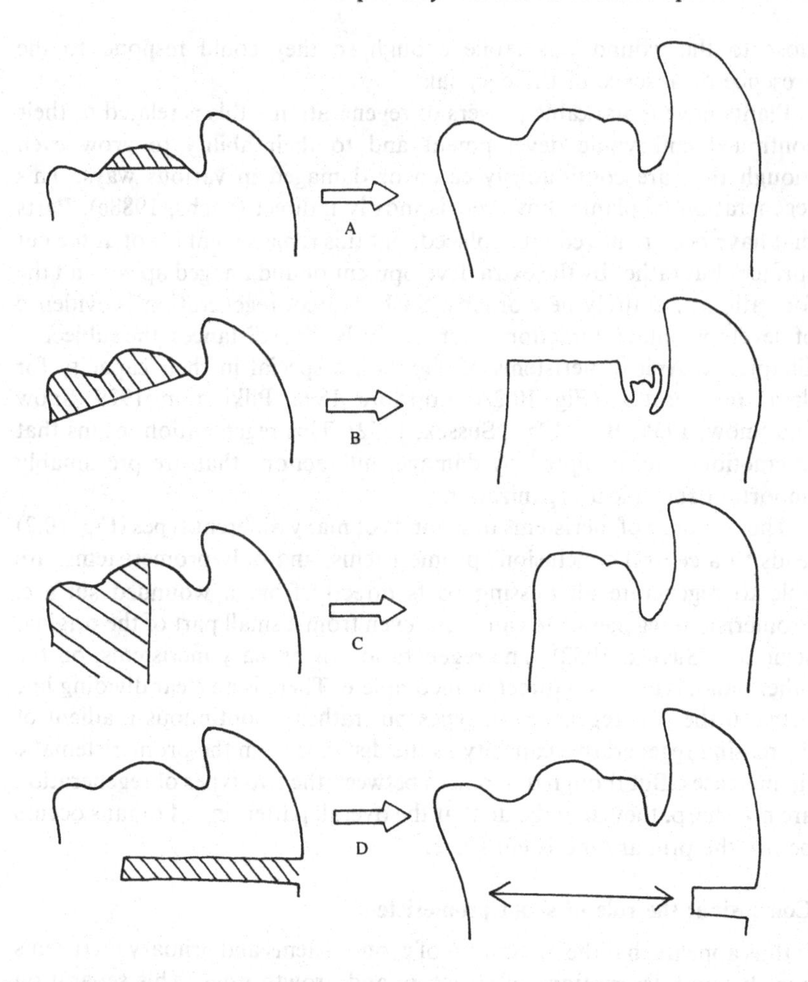

Figure 10.2. The capacity of shoot apices to regenerate after various wounds. Results based on literature reports on various species and on experiments on the lateral bud apices of pea seedlings. The apical outlines on the left indicate how the apices were wounded; those on the right the form of the regenerated apex, after growth had been resumed. A. Part of the dome above the leaves was removed. The missing parts were replaced directly from the wounded region. The apex regenerated leaving no sign of the original damage. B. The apical dome was removed together with a small part of the region from which leaves develop. Apices which underwent such operations regenerated only indirectly – by the growth of one or more lateral buds. The remaining parts of the damaged apex did mature, but their development was always limited since the promeristem was not replaced. C. A longitudinal half of an apex was removed. The apex continued to develop, but the complete replacement of the missing parts was limited to the promeristematic tip. D. A thick transverse section was removed from the circumference of the stem, below a young leaf. Though regeneration was not complete, transverse growth of the stem tended to decrease the influence of the wound as the stem matured.

close to the wound was labile enough so they could respond to the presence or absence of these signals.

Plants have remarkable powers of regeneration – this is related to their continued embryonic development and to their ability to grow even though they are continuously eaten or damaged in various ways. This regeneration of plants, however, is mostly indirect (Sachs, 1988c). Parts that have been removed are replaced, but this replacement is not at the cut surfaces but rather by the extra development of undamaged apices and the formation of entirely new organs. Such indirect regeneration is evidence of developmental interactions over relatively large distances, the subject of Chapter 2. Apical meristems of shoots are special in their capacity for direct regeneration (Fig. 10.2A; Lopriore, 1895; Pilkington, 1929; Snow and Snow, 1935; Ball, 1952; Sussex, 1952). This regeneration means that interactions are disrupted by damage, interactions that are presumably important for apical organization.

The response of meristems to wounds of many different types (Fig. 10.2) leads to a central conclusion: promeristems, and only promeristems, are able to regenerate all missing parts directly from a wounded surface. Promeristem regeneration can occur even from a small part of the original structure (Sussex, 1952). The regeneration of primary meristems, on the other hand, is always indirect or incomplete. There is no clear dividing line between the two regeneration types but rather a continuous gradient of decreasing regenerative capacity as the distance from the promeristematic tip increases. But though differences between the two types of regeneration are not sharp, they do indicate that the overall patterning of organs occurs before the primary meristem stage.

Conclusion: the role of shoot promeristems

It thus appears that the separation of promeristems and primary meristems goes beyond descriptions of structure and growth rates. This separation defines an undifferentiated promeristem which is the location of organogenesis and of overall patterning. The initial patterning which is associated with very little growth in the small promeristems is followed by the large increase in size at the primary meristem stage. The primary meristems are also sites of 'fine-tunning' of patterns: the formation of specialized cell types and the determination of their relative locations (Chapter 8).

The separation between promeristems and primary meristems is useful even though it is not based on all-or-none criteria. It stresses the separation of organogenesis from rapid growth. This separation is the rule even in the meristematic tips of plants with apical cells (Chapter 7), otherwise quite different from the apical meristems of seed plants. A reason for this separation between promeristems and primary meristems could be that organization processes can occur only over small distances

(Wolpert, 1971), and this precludes rapid growth. Another possible role of the slow growth of the organogenic promeristems may relate to their serving as a source of cells over long periods. Rapid divisions of these cells may not be consistent with their unchanged preservation (Barlow, 1978; Lerman, 1978; Klekowski and Kazarinova-Fukshansky, 1984a; Klekowski, 1988). Both these possibilities are reasonable and they complement rather than contradict one another.

THE ORGANIZATION OF ROOT APICES

Root apices resemble shoot apices in being centers of continued, potentially unlimited development. And the development of roots resembles that of shoots: the kinds of cells found in both mature tissues and the apical meristems themselves are similar if not identical. Both types of apices have the same general shape – an elongated axis with a dome at the tip. But the apices form different organs: roots have no leaves, their vascular tissues are arranged in a distinctive pattern in the center and there is a root cap distal to the apical meristem itself (Fig. 10.1B). A comparison of the apices of the roots with those of shoots could therefore contribute to a discussion of the organization of apices and, specifically, to an understanding of the relative roles of an organogenic promeristem followed by rapidly growing but determinate primary meristems. Concentrating on these aspects means ignoring important features special to root development, such as the role of the calyptrogen, the region that forms the root cap (Barlow, 1975).

Structural studies, growth rates and regeneration following wounds define a root promeristem with functions resembling those of the shoot promeristem. This promeristem is a small region situated just proximal to the root cap in which there are few or no cell divisions (Clowes, 1959, 1961, 1984; Němec, 1966; Barlow, 1984). As with shoots, only the extreme tip, about 0.1 mm in height below the root cap, undergoes complete and direct regeneration following either transverse or longitudinal cuts (Fig. 10.3A; Prantl, 1874; Němec, 1905; Kadej, 1970; Barlow and Sargent, 1978; Barlow, 1981; Rost and Jones, 1988). Though the growth of this region is at best very slow, it is maintained in a meristematic state, competent of renewed growth, even under adverse conditions, when all other tissues mature (Wilcox, 1954; Barlow, 1984).

Almost all growth, in both volume and cell numbers, occurs in the primary meristematic regions, where the overall patterning of different cell types can be distinguished. But in these regions of rapid growth the regeneration of new roots occurs only indirectly, through the formation of new meristematic apices (Rost and Jones, 1988). Isolation in culture has also demonstrated the determinate nature of the tissues right behind the tip (Reinhard, 1953). Thus the root resembles the shoot in having an apex

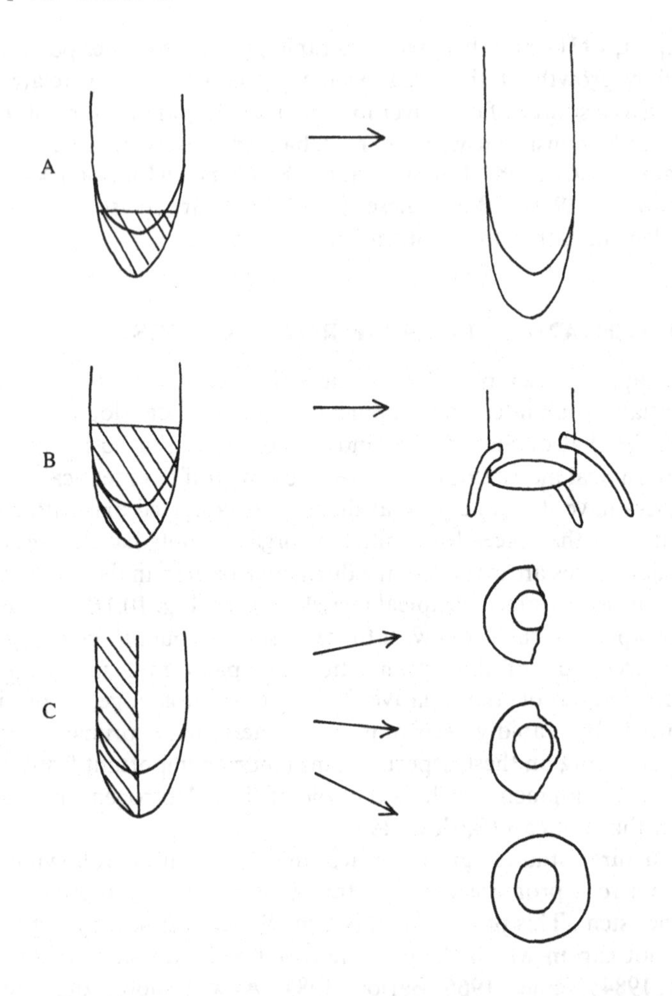

Figure 10.3. The regeneration of root apices after various wounds. Results based on literature reports for different plants and on experiments on roots of pea seedlings. Diagrams on the left indicate the way the apices were damaged and the ones on the right the form of the root after regeneration had occurred. A. The very tip of the root was removed, including the cap and the promeristematic tip. Such roots regenerated completely, the missing tissues being replaced from the surface of the wound. B. The cut removed a somewhat larger region, including less than 0.5 mm of the root behind the cap. The remaining tissues of such roots grew and matured, but the continuation of root development depended on indirect regeneration – on the replacement of the root tip by one or more lateral roots. C. A longitudinal half of the root was removed. The form of the regenerated roots is shown by cross-sections, the inner circle in each case representing the form of the vascular cylinder. Only the very tip of the root regenerated completely, into a circular root. In the tissues behind the tip, regeneration was always partial, decreasing gradually as the distance from the tip increased. The vascular cylinder always regenerated more completely than the surrounding cortical tissues.

consisting of a small, slow growing tip where tissue organization is determined – a promeristem – and much larger regions where practically all tissue growth and observable differentiation take place.

Yet the promeristems of roots are different from those of shoots. Root meristems are distinguished by a 'quiescent center' (Clowes, 1959, 1961, 1984; Steeves and Sussex, 1972; Barlow, 1976, 1984) – a region in the very tip which may not grow at all. Evidence for the absence of any cell divisions includes autoradiographic studies of apices supplied with DNA precursors, which show an absence of DNA synthesis in the quiescent centers (Clowes, 1961). And yet the quiescent center is the active source of cells during root initiation and when roots regenerate following damage, including non-local damage caused by cold (Barlow and Rathfelder, 1985), or even following an imposed dormant period (Barlow, 1976, 1984). It thus appears that the quiescent center could be the promeristem of roots, or a major part of this promeristem, even though it is not always an active, functioning promeristem. This suggests that in roots, in which no leaves are formed, existing tissues may expand as primary meristems for long periods and promeristematic growth is not always essential.

Though the subject has not received much attention, the presence of the quiescent center may influence the apparent number of initial cells in roots (Chapter 7). The evidence of chimeras concerning initial cell number is meager and conflicting (Sinnott, 1960). While the promeristem is quiescent, it could be expected that a large number of cells at its upper border would act as initials (Barlow, 1976). When the quiescent center is active a small number of initials could be expected, and perhaps even a single cell is the ultimate source of all tissues. The presence of the quiescent center may also be related to the size of root promeristems: though they are normally the same general size as shoot promeristems, they can also be many times larger. These large sizes are common in monocots, especially ones with large storage organs, and in large grasses such as maize.

POSSIBLE CONTROLS OF PROMERISTEM ORGANIZATION

The search for sub-regions

A possible approach to understanding the processes responsible for the organization of promeristems is to seek sub-regions within promeristems that could have definable roles and interact with one another. Some attempts in this direction were mentioned in Chapter 7: the separation of promeristems into a tunica and a corpus which reflect the orientation of cell divisions (Gifford, 1954; Clowes, 1961; Cutter, 1965; Steeves and Sussex, 1972; Esau, 1977; Fahn, 1982). These regions have important consequences for cell lineages (Chapter 7), but they are not known to be important for the organization of promeristems. Indications that no such organizational role should be expected is that the number of tunica layers is not constant, not even in the very same plant, and this number is not related to the organogenic activities of promeristems.

A further division of apices into 'cytohistological zones' was first described for gymnosperms (Foster, 1941) but later for angiosperms as well, especially ones with large apices (Gifford, 1954; Gifford and Corson, 1971; Steeves and Sussex, 1972; Mauseth, 1978). These zones are based primarily on the orientations of cell divisions, but there is also evidence for biochemical differences between the cells and for quantitative differences of the distribution of cellular organelles (Mauseth, 1981). These zones, however, are not general, they vary greatly even in a given plant (Gifford and Tepper, 1962; Mauseth, 1978) and their significance for the organization of meristems, if any, is not known.

The possible role of interactions with the plant

The approach followed here is that an entire promeristem could be a functional unit. This is at best a rough approximation, but it may be the right way to start in view of the absence of concrete evidence for interacting zones. This suggests considering the interactions of the promeristems with those of the rest of the apical region and with the entire plant (Sachs, 1972b; Barlow, 1976). Though it was concluded above that meristems do not receive detailed or patterned information, *directional* interactions that limit growth and occur through the base of the meristem were not ruled out – such interactions are possible even with unorganized callus and with a culture medium. There is no doubt that there are such interactions: factors necessary for the development of promeristems or for the control of its activities, which might well include plant hormones, must enter through the base of the promeristem. Furthermore, substances that serve as signals that coordinate promeristem activity with the rest of the plant must pass through the same basal cells.

What could be the 'limiting supplies' and 'developmental signals' for the development of promeristems? They could not be sugars or other building blocks of all metabolic activity: these are not unique to promeristems and are required by promeristems in relatively small quantities. There are some indications that the limiting supplies could be hormones, even known hormones (Sachs, 1972b, 1975b; Barlow, 1976, 1984). Cytokinins are known to induce shoot promeristem formation and to be essential for their continued activity (Chapter 3). Furthermore when cytokinins are present in excess quantities they disrupt organization – they cause teratomata development (Chapter 4). Cytokinins applied directly to shoot apices also have disruptive effects on organization (Wardlaw and Mitra, 1957; Sachs, unpublished). Root meristems are induced by auxin

(Chapter 3) while the addition of excess auxin disrupts the structure of root apices.

Thus though there is no reason to assume that auxins and cytokinins, or other known hormones, are the sole controlling factors of promeristem organization, there are facts that suggest that they are components of such controls. Limitations of other substances, including metabolites, are likely to be important. Another possibility is the physical strains that must arise during growth. Since these strains could act on the orientation of the cellulose microfibrils, and the cellulose microfibrils in turn orient and withstand continued strains, there could be feedback relations that would generate form (Green, 1980, 1988; Barlow, 1984; Lintilhac, 1984). This general possibility is considered in somewhat greater detail in relation to phyllotaxis, in the following chapter.

An hypothesis concerning the control of promeristem organization

Gradients and flows of substances through the base of promeristem could induce the polar, determinate differentiation of the cells leaving the promeristematic region (Chapter 5; Barlow, 1976). Such inductive interactions between the promeristems and the plant could thus determine the height of promeristems: the cells at the tip of the shoot or the root would not be induced to polar differentiation because their very location would preclude the occurrence of gradients and flows of controlling substances. As cells are added to promeristems by division, their activities as sources and sinks of controlling substances could lead to the differentiation of cells at the base. Thus the overall height of the promeristem would remain constant while its growth continues.

This raises the question whether controls of the supply of limiting factors could be responsible for other aspects of promeristem organization. Major facts that these controls would have to account for are the maintenance of the size and form of promeristems, which require quantitative relations between the growth in various regions and directions (Hejnowicz, 1955). It is also necessary to account for the increase of promeristem size following initiation, the remarkable regenerative capacity of promeristems and the common inhibition by formed promeristems of the initiation of new, adventitious promeristems.

These traits of promeristems suggest a positive feedback control of promeristem development and organization (Fig. 10.4; Sachs, 1972b; Lyndon, 1979), an extension of the relations of entire apical meristems considered in Chapter 2. The development of promeristems could be limited by the signals it receives from the rest of the plant. The supply of these limiting signals to a promeristem could in turn depend on the promeristem being a source of complementary, inductive signals that influence basal differentiation and channel the limiting factors from the

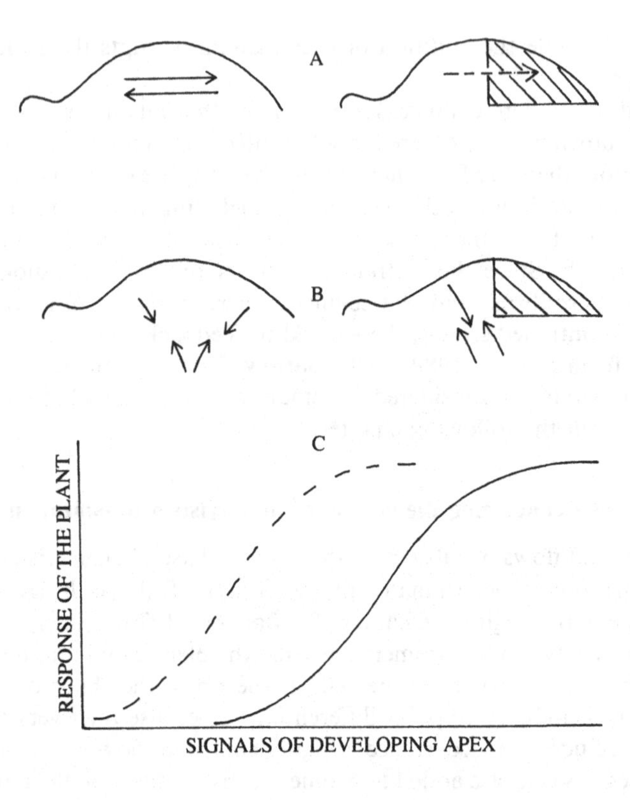

Figure 10.4. Hypotheses concerning promeristem organization. A. Transverse interaction across the promeristem. The removal of one half leads to reduced inhibitory stimuli – to the absence of the dotted arrow – reaching the remaining half; growth is increased until the original size and form are restored. B. An alternative which does not require direct interactions across the promeristem. The development of the various parts of the promeristem depends on limiting signals they are abler to divert towards themselves. The removal of half the promeristem increases the diversion of signals towards the remaining parts, this leading to increased growth until the original size and the round shape are restored. C. Representation of the quantitative aspects of the suggestion diagrammed in (B). There is a non-linear relation between the signals of a developing apex and the growth-limiting responses they elicit from the rest of the plant. Both the signals and the response can be measured by the rate of promeristematic development to which they are related. Signals of few cells elicit a weak response from the plant. Signals from half a promeristem elicit a large response, sufficient for an increase in the size of the promeristem. This cooperative response to signals of neighboring cells does not continue indefinitely – an intact promeristem is able to maintain but not to increase its size. The entire curve is shifted (broken line) in plants in which there is no competition from other promeristems and the limiting 'responses' of the plant are present in excess.

plant. At the early stages of promeristem initiation the channelling signals could be the mere use of limiting factors by the promeristematic cells; this would result in diffusion towards the new promeristem. Later, the oriented differentiation of the cells at the base (Chapters 6, 7), and perhaps also additional, long-distance directional effects of hormones (Patrick, 1976) could be added as major diverting factors.

If this is true, the larger the promeristem and the higher its developmental rate, the more signals influencing the rest of the plant it would produce and the more limiting supplies it would receive. This relation could determine form if there were no simple linear relation between the development of promeristems and the limiting signals they divert towards themselves (Fig. 10.4B,C). Single cells or small centers may receive no supplies, while promeristems at half their final size may divert excess supplies, sufficient for continued enlargement. Such a positive feedback involving supply and demand would have an upper limit, a state beyond which increased development does not elicit increased supplies. Promeristems would then develop until a balanced supply and demand state is reached. The same would occur during regeneration, resulting in the restoration of the original size of the promeristem. Since isolated cells would be at a disadvantage in obtaining limiting signals for further development, the round, domed shape of promeristems would be maintained.

The suggested controls lead to some predictions that are readily confirmed. The size of promeristems could be expected to depend on the availability of limiting substances – on the vigour of the plants, as indeed it often does (Cutter, 1955, 1965). This relation of meristem size with plant vigor is most clearly seen in roots of monocots, such as maize. New shoot promeristems are formed at the base of young leaves – where there could be an excess of supplies as well as of cells competent to respond to these supplies by initiating the feedback relations that lead to promeristem organization. In roots, new promeristems arise where there are active shoots and where other roots, which should act as competing sinks for the signals of the shoots, are not too close (Charlton, 1982). Excess supplies also occur on plants from which all comparable meristems have been removed – where there is no competition from organized promeristems. In such plants, if there are competent cells, new, adventitious apices are formed. Excess supplies can also be the cause of the unusual structure of 'fasciated' apices that form on vigorous plants, especially when all competing apical meristems have been removed (Lopriore, 1904; Gorter, 1965).

An outstanding characteristic of the suggested control is that it shows how cells could collaborate, acting as a unit which diverts more limiting factors towards itself than the sum of the effects of individual cells. It is remarkable, though perhaps counter-intuitive, that transverse dimensions could be controlled without transverse interactions (Fig. 10.4B). It is of course likely that such local transverse interactions do occur; the hypothesis outlined above only raises the question whether they need be specific and complex. Perhaps the clearest evidence for transverse interactions is found in the changing patterns of the vascular tissues of roots. The number of xylem poles, as seen in cross-section, varies during development and can be manipulated in damaged and cultured roots (Torrey, 1955, 1957; Feldman, 1977, 1984; Lyndon, 1979; Barlow, 1984). These patterns, however, are closely correlated with the size of the promeristem; they can be understood as resulting from the different 'packing' of units of a determined size (Chapters 8, 11) within promeristems of variable sizes (Sachs, 1981a). The size of the units of vascular tissues may result from interactions between the phloem and the xylem (Chapter 6); additional transverse interactions may occur, but they are not required by the available evidence.

THE DETERMINATE NATURE OF PRIMARY MERISTEMS

The primary meristems are the site of almost all apical growth. It is in them that the patterns outlined as the cells leave the promeristems acquire the rich cellular details found in mature tissues. Yet an outstanding characteristic of the developmental processes and developmental capacities of primary meristems is that they are 'determined'. This concept of determination requires an operational definition and a consideration of its possible basis.

The obvious expression of determination is that the development of primary meristems is always limited. It is true, of course, that the rate and duration of this development are readily influenced by environmental conditions. An outstanding example is the influence of light: in the absence of light the stem internodes are 'etiolated' and may grow to several times their length in strong light. This is due to the duration of growth in the primary meristems and not to any change in the promeristems (Thomson and Miller, 1963). Primary meristematic growth is also influenced by correlative effects of the rest of the plant, especially the inhibitory effects of similar organs (Chapter 2). Yet even when there are no competing organs the organized development of primary meristems eventually stops and can continue only where new primary meristematic tissues are formed by promeristems. The determinate nature of primary meristematic development is also seen in limitations of direct regeneration, mentioned above. Wounds are covered and closed readily and the continuity of the vascular system is regenerated (Chapter 5), but the former outlines and organs such as leaves are not restored (Figs. 10.2, 10.3).

The determinate nature of primary meristematic growth is not due to a limited supply of any essential factors. Primary meristem development stops even with an unlimited supply of nutrients from storage organs and where no other meristems are available for its use. No hormonal additions to intact or wounded plants are known to prevent the maturation of the primary meristematic tissues (Klein and Weisel, 1964). The impossibility of extending development with external supplies is confirmed in culture conditions: there is no problem in growing tissues as unorganized callus, which has no defined axis of growth (Chapter 4), but otherwise continued development requires the initiation of promeristems and the continued formation of new primary meristems. One suggestion has been that tissue maturation is due to a gradual dilution of the plasmodesmatal connections between neighboring cells (Gunning, 1978; Lyndon, 1979). However, new plasmodesmata do form across non-dividing walls (Chapter 9) and tissue maturation is not always coupled with growth.

It follows that the limitation of the development of primary meristems, their determinate nature, must be a function of the developmental processes within the meristematic tissues (Lyndon, 1979). The initial events may well be induced by the gradients (Barlow, 1976) or flows of inductive substances (Chapter 6) associated with the non-terminal position, but continued events may be governed within the meristematic cells themselves. There is no lack of evidence that this development leads to changes in the competence of tissues to respond to environmental and hormonal influences (Meins, 1986; Meins and Wenzler, 1986; Wenzler and Meins, 1986). It is thus not far-fetched to suggest that maturation itself is but a final stage of these changes, causing primary meristem development to be determinate or limited under all known conditions. The developmental changes are generally gradual, but there is no known way to stop them or even to slow them down. When growth is slowed down by environmental conditions the changes towards the mature state may continue and growth that did not occur is often 'lost' – it can only be made up by new, younger tissues. Maturation is thus a function of gradual change in the potential for growth, while the realization of this potential is influenced by many factors, including light and the correlative inhibition of other organs.

THE DIFFERENTIATION OF APICAL MERISTEMS

Plants have various types of shoots and, though this is less obvious, also various roots (Barlow, 1986). Comparative plant morphology showed long ago that these can be viewed as modifications of a small number of master organs or categories. From an ontogenetic point of view these facts suggest that apices share some underlying early processes and they vary in the relative role of various components of their development (Sachs, 1982). There are exceptional organs that cannot be reasonably placed within any category. There are also many 'intermediate' organs that share traits of different categories (Sattler, 1966, 1974). These can be understood as results of the combination of primordial processes from different apical types in various ways (Corner, 1958; Sachs, 1982).

This means that apical meristems can have many differentiated states – for structures of varied nature are necessarily the products of different developmental processes. The discussion above would suggest that differentiation could occur in either the promeristems or in various processes in the primary meristems. It also suggests that a comparison of the differentiated apices and a definition of the processes that vary while others remain unchanged could be interesting in itself and could contribute to an understanding of apical meristems. An outstanding example of apical differentiation are the differences between shoot and root apices which were mentioned in the discussion above – and these differences could be due to different specializations of the respective promeristems for the requirement and production of limited hormonal signals (Wareing, 1978).

The special characteristics of the various vegetative shoots, often found on the very same plant, involve quantitative variations of processes common to most if not all primary meristems. For example, potato tubers are produced at the ends of stolons with apical meristems that differ from those of the leaf-bearing shoot in a lower rate of cell divisions and more rapid cell elongation (Clowes and MacDonald, 1987). The tubers themselves have 'normal' meristems and all organs of shoots and the distribution of the tissues and the types of cells are also usual. The form of tubers is a result only of an exaggerated transverse growth of the ground meristem. Examples of other modifications are brachyblasts, which are special in having very short stem internodes, thorns which are formed by meristems in which lignification is unusually pronounced, and tendrils which have extended intercalary growth. In both thorns and tendrils the activity of the promeristem stops at an early stage – but little is known about this, though some descriptions are available (Millington, 1966; Bienick and Millington, 1967; Tucker and Hoefert, 1968; Steeves and Sussex, 1972). Since there are many types of apical differentiation, there are no valid generalizations about the locations or types of processes that vary and a full discussion of the differentiation of apices would be little more than a long catalogue of incomplete information. For this reason only reproductive apical meristems, which are an example of major differentiation events in the promeristems, were chosen for discussion below.

Reproductive apices

A shoot apex may form various numbers of leaves and then change to form a flower or an inflorescence (Chapter 12). Morphologically the flower is a modified shoot – and this means that reproductive apices are not only formed from vegetative apices, they also share basic developmental processes (Sachs, 1982). Thus the formation of flowers is a major expression of an apical differentiation common to all seed plants.

So as to consider the processes of differentiation during flower formation it is necessary to focus on a few major apical changes. The stress here will be on two related aspects: the stepwise change in the character of the lateral organs, and the determinate or limited continuation of embryonic development. These are events that occur within apices and do not include the most studied topics related to flowering: the conditions external to the apex that induce the transition to reproductive development (Bernier et al., 1981; Kinet et al., 1985). The effects of the environment are outside the scope of this book (Chapter 1) and the internal control by the age or size of the plant is the subject of Chapter 12.

A first question concerning flower formation may be whether changes occur in the promeristems themselves or only in the primary meristems below. Sections through apices undergoing the transition to reproductive development show anatomical and cytochemical changes (Bernier et al., 1981). A general metabolic activation and a general enlargement of the promeristematic region are common and any zonation of the cells of the vegetative apex disappears during the transition to reproductive development. But the increase in promeristem size during flower formation is not the rule for all plants (Lyndon, 1977; Francis and Lyndon, 1979) and could therefore not be an essential component of the apical changes. An alternative change may be a reduction of the initial size of the lateral organs, as compared with the leaves on a vegetative apex (Lyndon and Battey, 1985). Promeristematic changes are thus common, but it is not yet clear whether they are necessary components of the transition of apices to the reproductive state.

The different types of lateral organs follow one another in rapid succession (Sattler, 1973). The appearance of these organs indicates the order in which they are determined and studies of the regeneration of reproductive apices confirm the sequential determination of the lateral organs (Cusick, 1956; Soetiarto and Ball, 1969; Hicks and Sussex, 1971). The reproductive apex behaves as a promeristem: only organs that were not present at the time it was split or damaged are formed all around a reconstituted apex. Organs that had already appeared are not replaced. Thus an important trait of reproductive apices is that developmental stages, characterized by the nature of the lateral organs, are not repeated once they have been achieved. In spite of this, the determination of an entire apex can be changed and even reversed. Such changes occur in some plants when the environmental conditions that induced flowering are experimentally changed in unnatural ways (Battey and Lyndon, 1986). Changes are also caused by homeotic mutations that increase the number

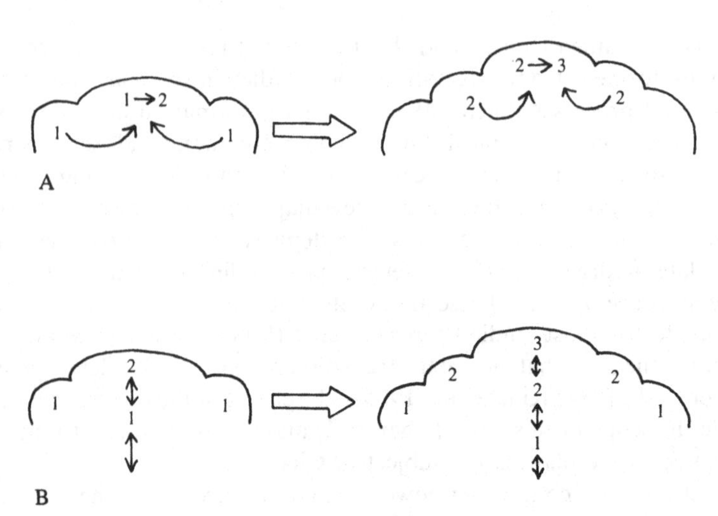

Figure 10.5. Two hypotheses concerning developmental changes in reproductive apices. The problem is the cause of the transitions from the formation of sepals to petals, stamens and carpels. A. The primordia of organs that have already been formed influence the apical dome. their influence causes a change in the apex – from 'phase 1' to 'phase 2'. The new primordia will cause a further change, until all parts of the flower have been formed. B. The apex changes as a result of the accumulated 'physiological distance' from the rest of the plant (Chapter 12). The lateral organs would be expressions but not causes of apical changes.

of organs of one type by changing the primordia of the following type: such as the change of a stamen into a petal. Examples of mutations that change the number and fate of organs are common in garden plants, where various increases in the number of petals have been successfully selected for in many species, roses being the most obvious example. Genes with homeotic effects on flower development are now being identified and isolated, especially in *Arabidopsis* (Pruitt et al., 1987).

The sequential changes in the reproductive apical meristem have been interpreted as meaning that each type of lateral organ induces the formation of the next lateral organs – sepals induce petals, etc. (Fig. 10.5A; Heslop-Harrison, 1963). The inductive effects end when the formation of the carpels leads to the cessation of all promeristematic development. Direct tests of this hypothesis by damage of all lateral organs of a given type, at the time they first appear, have not been reported. Another interpretation of the available evidence is that the promeristem itself undergoes gradual changes, expressed but not caused by the stepwise changes in the lateral organs (Fig. 10.5B). This suggestion is attractive because it relates directly to a process known to occur in apical meristems from other, independent, evidence (Chapter 12): the transitions from juvenility to maturity show that a gradual accumulation of changes in promeristems can actually occur. As a working hypothesis it is reasonable to assume that the changes typical of reproductive apices are the ones common to all plants and that these same processes have been extended over longer periods, more suitable for experimental manipulations, as evolutionary responses to the need for juvenile stages. The discussion of sequential apical differentiation is taken up again in Chapter 12.

11

The localization of new leaves

The arrangement of leaves on stems, the 'phyllotaxis', is orderly: this is expressed by the predictability of the location of a leaf from the location of other leaves (Figs 11.1, 11.2). There are various phyllotactic patterns (Green and Baxter, 1987), and these depend on the species and often also on the stage of plant development and environmental conditions. Phyllotactic patterns are prominent, macroscopic and obvious in mature tissues, and they have remarkable mathematical properties. For these reasons they have received much attention, perhaps as much as any other biological pattern.

Most work on phyllotaxis has dealt with the mathematical or geometrical aspects of leaf arrangement (Church, 1904; Iterson, 1907; Richards, 1951; Mitchison, 1977; Erickson, 1983; Jean, 1984). These have been especially concerned with the mathematical properties of the different angles of divergence between leaves and of the lines connecting neighboring leaves (the parastichies), but this aspect of phyllotaxis is not directly related to the purpose of this book (Chapter 1). There have also been quantitative models of phyllotaxis that have supported detailed mechanisms concerning the relations between leaves – but these models have been based on unrealistic assumptions which bear no relation to actual plant development: assumptions that leaves of seed plants are not in direct contact at the time they are formed, that their bases are round and that these bases are passive, unchanging participants in apical events.

The discussion in this Chapter will therefore center on experimental work concerning an early and major aspect of phyllotaxis: the localization of the initiation of leaves on the shoot apex (Fig. 11.2). This initiation can be specified in terms of the angle between neighboring leaves and the number of leaves at each node. Although leaf initiation is the major organogenic process in the shoot apex, the question of its localization has not received much attention, especially from the experimental point of view. Thus the following discussion will ignore much of the literature on phyllotaxis and stress the work of three laboratories, those of the Snows, Wardlaw and Green. Reviews that include other aspects of the subject are:

Figure 11.1. Phyllotactic patterns. The youngest leaves are always above, and thus inside, the previous leaves. A. *Graptopetallum*. An example of a spiral phyllotaxis, in which the leaves are formed one at a time. B. *Crassula*. Leaves are formed two at a time, on opposite sides of the apex. C. *Kalanchoë*. Leaves in whorls of three. D. *Vinca*. Leaves in pairs, as in (B), but in this plant and in other climbers the elongation of the internodes separates the leaves very early, before they reach their final size.

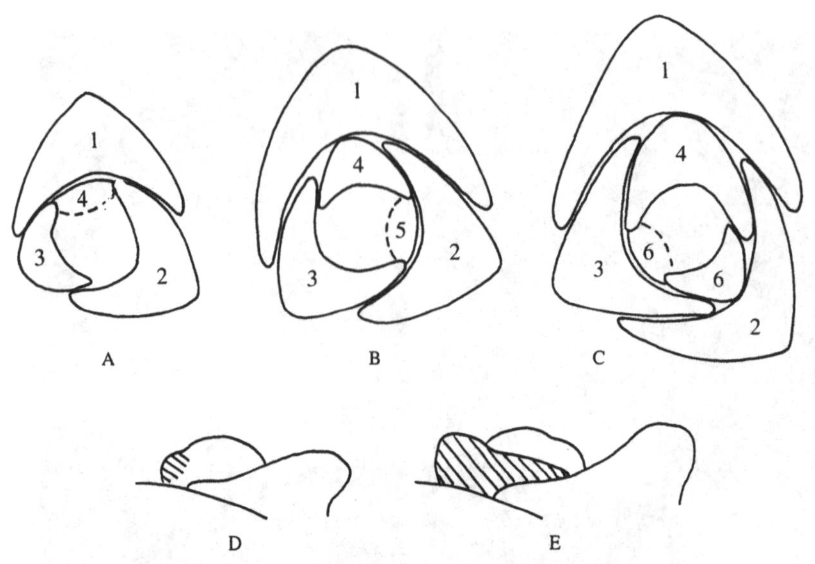

Figure 11.2. The localization and the spreading of new leaves. A–C. The drawings indicate consecutive stages in the development of the same shoot apex, as seen from above. This apex is surrounded by leaf primordia. As the apex grows, new competent tissues become available and new leaves form in the first large enough space on the shoot apex. This space is always at the greatest possible distance from the previous leaves. As the apex grows further the new leaf spreads above the previous leaves, eventually surrounding half or more of the apical circumference. This spreading has a profound effect on the location of the next large enough space for leaf initiation. D–E. The same leaf initiation and spreading, viewed from the side of an exposed apex. The tissues competent to form new leaves are always above the previous leaves, just outside the apical dome.

Snow (1955); Sinnott (1960); Cutter (1965); Steeves and Sussex (1972); Williams (1974); Rutishauser (1981); and Schwabe (1984).

THE LOCATION OF NEW LEAVES IS DETERMINED WITHIN THE APEX

The initiation of new leaves may reflect early events that are not readily observable. These events could be localized by signals from outside the apex and by processes within the developing apex itself. Since both specification from outside and from within the apex could be important, it is necessary to seek evidence concerning the relative roles of these two possibilities.

The environment does not supply any information that has sufficient detail to determine the location of a leaf (Chapter 1). Any localizing influence from outside the apex would therefore have to come from the relatively mature tissues of the shoot. Such influences could not be the sole

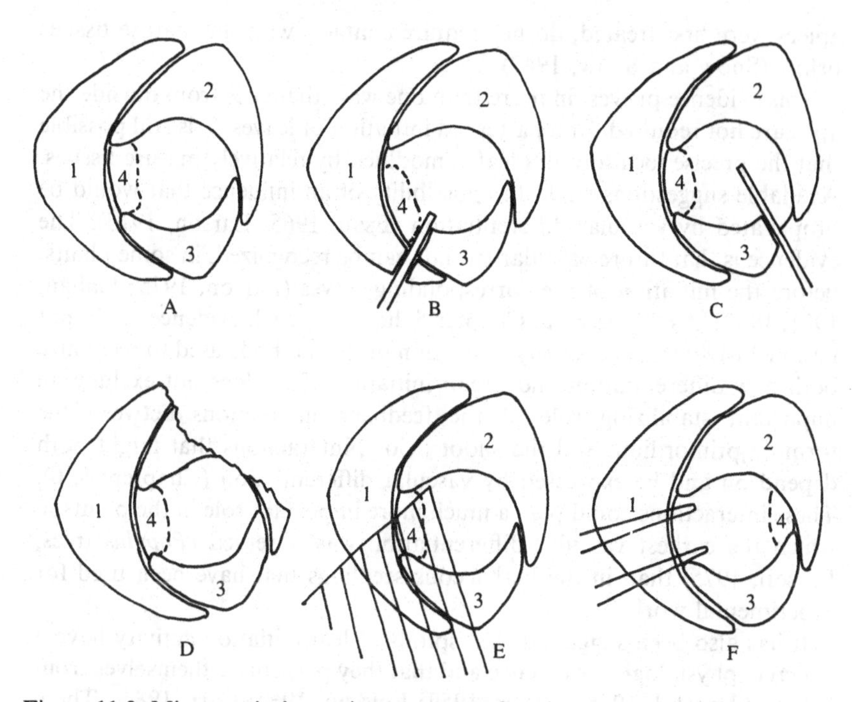

Figure 11.3. Microsurgical experiments concerning the localization of new leaves (based on the publications of Snow and Snow). A. An untreated apex; the new, fourth leaf is forming in the largest space above the previous leaves. B. The apex was cut; this interfered with the spreading of the third leaf primordium and shifted the location of the fourth leaf. C. A similar cut, but one which did not impinge on the space of the future fourth leaf. Such cuts did not change the location of the fourth leaf. D. An existing leaf primordium was severely damaged. This damage should have influenced any inhibitory effects this primordium had, but it did not change the location of the next leaf primordium. E. The triangle represents a transverse cut, isolating the base of the future leaf primordium from the plant below. Such cuts neither prevented nor modified leaf initiation. This was true even when the location of the new leaf could be shifted by microsurgical treatments, as in (B), so their location could not have been determined at the time the apices were treated. F. Two cuts isolated part of the apex from the effects of existing leaves. New primordia did not form if the space between these cuts was small.

determinants or phyllotaxis: apices precede the shoots they form, and a phyllotaxis normal for the species appears even when the apices are initiated on callus rather than an organized plant. Further evidence comes from microsurgical experiments (Fig. 11.3). Shallow cuts, limited to the tip of the shoot apex, influence the location of future leaves (Snow and Snow, 1931, 1933). Undercutting the location of future primordia, on the other hand, does not prevent their appearance in the expected location (Wardlaw, 1956). Even primordia whose location could be changed by vertical cuts, and thus could not have been fully determined at the time the apices were first treated, do not require contacts with the mature tissues below (Snow and Snow, 1947).

This evidence proves, in more than one way, that cues from outside the apex are not required for a patterned initiation of leaves. It is still possible that the precise location of a leaf is modified by relatively mature tissues. Available suggestions stress the possibility of an influence that would be propagated by vascular differentiation (Esau, 1965; Larson, 1975). The evidence is that future vascular strands can be recognized, in some plants, before the initiation of the corresponding leaves (Larson, 1975; Gahan, 1981, 1988). As discussed in Chapter 5, however, such evidence could not be conclusive: it is necessarily a function of the methods used to recognize both new differentiation and organ initiation. This does not exclude an important stabilizing role of the feedback interactions between the forming primordium and the shoot below, interactions that might both depend on and be expressed by vascular differentiation (Chapters 5, 6). These interactions could play a much more important role in the plants in which the earliest vascular differentiation was observed (*Populus* trees, Larson, 1975) than in the herbaceous seedlings that have been used for experimental work.

It has also been suggested that spirals of leaf initiation activity have a concrete physiological existence and that they perpetuate themselves from below (Plantefol, 1948; Cutter, 1959; Loiseau, 1969; Carr, 1984). These spirals were supposed to vary in number, depending on the characteristics of the phyllotactic system. Microsurgical evidence for these spirals (Loiseau, 1969) can also be understood on the basis of controls that operate within the apex (Cutter, 1965). Leaves can appear, furthermore, in the space between two vertical cuts that would be expected to interfere with the propagation of the generative spirals (Fig. 11.3F; Wardlaw, 1950; Snow and Snow, 1952). Furthermore, any essential role of spirals that perpetuate themselves from below is ruled out by the evidence mentioned above, that the location of leaves is determined within the apex itself.

A simple description, at this stage not meant to imply any hypothesis, may be a good start for the discussion of the localization of leaves within an apex. Leaf primordia appear only on shoot apices, always right below the promeristematic tip (Fig. 11.2). Observations of leaf initiation show that the centers of primordia are always at the center of the first available space outside the promeristem and above previous primordia (Hofmeister, 1868; Iterson, 1907). Once the primordium is initiated, it spreads so that it includes all neighboring space (Fig. 11.2; Snow and Snow, 1962). In many plants the regions added after initiation are the origin of the stipules, though they may also develop as parts of the leaf proper. Irregular initiation occurs in *Acacia* species (Rutishauser and Sattler, 1986), but these are rare exceptions that could represent early stages in the evolution of new morphogenetic controls: they stress rather than contradict the principles that apply to almost all plants. Finally, there are reports of adventitious leaves that were formed in the absence of a shoot apex (Goebel, 1900). These reports are questionable, since a small apex may have been present and yet difficult to detect at the time the leaf appeared as a bulge. It is therefore not clear whether such leaf initiation without apices occurs, but even if it does it is certainly unusual.

THE LONGITUDINAL LOCALIZATION OF LEAF INITIATION

It is now useful to distinguish two components of the location of a new leaf (Fig. 11.2). The first is relative to the shoot tips and along the plant axis. The second is around the circumference of the axis, this being a major component of the relation of a new leaf to the other leaves. For reasons of clarity these two aspects will be considered separately, but it should not be forgotten that they are parts of the very same process. The first topic here will be the localization along the plant axis.

As stated above on descriptive grounds, leaf initiation occurs only right below the tips of shoot apices (Fig. 11.2). Cells outside this region form no leaves even when all possible inhibitory organs are removed. Nor do they form leaves directly, without first forming a new apex, when they are isolated by cuts or are grown in culture. To account for the absence of leaves in the promeristem itself it has been assumed that the tip inhibits leaf formation. But there is no reason to assume that the cells of the promeristem are able or competent to form leaves – so there is no need for the assumption that their formation is continuously inhibited.

The position of initiation right below the shoot tip suggests a possible positive or inductive influence of the tip. Information about this possibility can be found in microsurgical experiments. In ferns the removal of the shoot tip causes newly initiated leaves to become buds, which include an entire shoot apex (Wardlaw, 1950, 1968; Cutter, 1965). This indicates that a continued inductive effect of the shoot tip is required for the determination of a primordium as a leaf. In seed plants, on the other hand, the removal of the shoot tip does not change the fate of leaves that have already been initiated (Pellegrini, 1961). Leaves that appear after the apex has been damaged may have primordia of unusual sizes (Snow and Snow, 1955) and they may be abnormal, lacking lateral expansion ('Centric organs' or 'Radial leaves'). It is not clear whether this unusual from is due to an absence of an inductive influence of the promeristematic tip (Sussex, 1955) or to direct damage of the leaf primordium itself (Snow and Snow, 1959; Sachs, 1969b).

It can only be concluded that the determination of leaves occurs very early, close to the tip, before the primordium appears as a bulge of tissue.

The facts are perhaps best described as expressions of a special competence to form leaves in a non-terminal region that is very close to the shoot tip. A localized competence along the meristematic axis is in accord with other evidence concerning the structure of shoot apices (Chapter 10). The leaf-forming region is the 'anneau initial' of Plantefol (1948; Loiseau, 1969), except that this region does not require any unknown, self-perpetuating structure. The special traits of the cells that form leaves could be associated with their location: these are the cells that undergo the first changes from the promeristematic to the polar, rapidly growing state; these changes are a function of their non-terminal position (Chapter 10). Continued cell maturation leads to further changes that make leaf initiation impossible. This is, however, only a description or at best a working hypothesis that could be useful for further work. Although the nature of the competence to form leaves is not understood, the limitation of leaf formation to a narrow region, just below the apex, is necessary so as to account for the vertical component of the localization of leaf primordia.

THE CIRCUMFERENTIAL LOCALIZATION OF NEW LEAVES

As stated above, the availability of space on the circumference of the competent regions on an apex is correlated with the formation of new leaves. This correlation could be more than a descriptive fact, space determining the location of leaf initiation in the two following ways (Fig. 11.4). (a) Existing primordia could be the source of an influence that inhibits leaf initiation and this influence could decline as a function of distance. New primordia would then arise only when there is tissue at a sufficient distance from the primordia that are already present (Wardlaw, 1950, 1968). The inhibitory influence could be due to the leaf serving as a sink for an essential resource or to the actual formation of inhibitory signals by the new leaves. (b) A new primordium may require a minimal space or volume to become organized, and only when the apex has grown sufficiently does a large enough region become available. This entire region would be required for the formation of a leaf because a leaf is a multicellular entity from its very inception (Chapter 7; Iterson, 1907; Snow and Snow, 1933, 1947; Snow, 1955).

These two possibilities that would account for the need for space for leaf initiation are not mutually exclusive and both are in accord with the descriptive facts of phyllotaxis. From a mathematical point of view they might well be identical. The two possibilities do differ, however, in the physiological control mechanisms they would require: one predicts the existence of short-range inhibitory interactions and the other the cooperation of many cells in the formation of an organized entity, a leaf

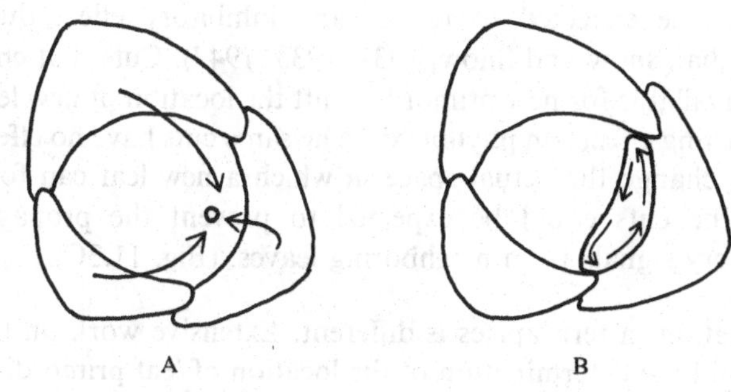

Figure 11.4. Two hypotheses concerning the space required for the initiation of new leaves. A. The center of the new leaf must be at a sufficient distance from existing leaf primordia. This distance is required so inhibitory effects should become sufficiently weak to allow new initiation. B. The primordium is first organized as a multicellular structure, not as a specialized cell or point. Space above previous leaves must be large enough for this organization to take place and the previous leaves need not be sources of any inhibitory influence on future leaf initiation.

primordium. The existence of inhibitory interactions has often been taken for granted, especially in mathematical models. There is, however, some direct evidence concerning the nature of the requirement for space. The indications differ for seed plants and for ferns, and they will be considered separately.

In seed plants, the following descriptive and experimental facts suggest that the organization of a new leaf require a minimal, multicellular space.

(a) The growth of the shoot apex is always associated with leaf initiation. Leaf primordia occupy the entire available space of competent tissue in an apex; where there are distances between leaves they appear only later in ontogeny, after the determination of leaves. Thus, all the cells of the shoot tip that are competent to take part in leaf formation do so, regardless of their distance from neighboring primordia.
(b) The first observable events in the formation of a new leaf are always multicellular. Thus, the reorientation of growth and cell division, leading to the appearance of a leaf as a bulge, are not limited to any single cell (Chapter 7; Lyndon, 1982). The reorientation of cellulose microfibrils, considered further below, also spans many cells (Green, 1985, 1986).
(c) Leaf initiation following microsurgery also provides evidence that many cells must cooperate in the initiation of a new leaf (Fig. 11.3F). Leaves do not arise between two cuts that are close together, even though these cuts isolate the intervening tissue from inhibitory effects of neighboring leaves (Snow and Snow, 1952). Damage to the leaf primordia does not change the location of future leaves even though

> it could be expected to reduce any inhibitory effect due to the primordia (Snow and Snow, 1931, 1933, 1947). Cuts that change the space available for new primordia shift the location of new leaves and have lasting effects on phyllotaxis. The same cuts have no effect if they do not change the actual space in which a new leaf can form, even when the cuts could be expected to prevent the propagation of inhibitory signals from neighboring leaves. (Fig. 11.3C)

The situation in fern apices is different. Extensive work on the apical structure and the determination of the location of leaf primordia in ferns has been carried out by Wardlaw and his students (Wardlaw, 1950, 1956, 1968: Cutter, 1956, 1965). Leaves are initiated as a single specialized cell that is not in direct contact with neighboring primordia. Most of the surface of the fern shoot apices is not occupied by leaf primordia at any stage. Microsurgical cuts influence the location of neighboring leaf primordia, as they do in seed plants. New primordia originate, however, in the space between two cuts that are much closer than neighboring leaves on an intact apex (Wardlaw, 1950).

It thus appears that the localization of leaf primordia may not depend on only one major process that is equally important in all plants. There is evidence of a requirement for both a minimal area or volume of tissue for leaf initiation and inhibitory signals of young leaf primordia. The relative role of these processes differs depending on the group of plants that is studied; in seed plants the need for a minimal area appears to be the dominant factor. The experimental evidence is convincing, but it all comes from limited numbers of plants and primarily from one laboratory. Some of these experiments have been repeated (Sachs, unpublished), but additional work is necessary before any firm conclusions can be reached.

CONCLUSION; LEAF INITIATION AND PHYLLOTAXIS

The available facts suggest that in seed plants the relation of the 'first available space' to initiation is not only a description of apical development; it also determines the localization of new leaves. A primordium becomes organized wherever competent cells are present in a sufficiently large mass. Competent cells are continuously formed by the growing apex and the occupation of space by existing leaf primordia results in the center of a new leaf being formed at the greatest possible distance from the centers of older leaves. The growth of the apex and the occupation of available space by new leaves result in a self-perpetuating, periodic process of orderly leaf initiation. This stresses the importance of cell competence and the supracellular nature of organ development (Chapter 7) for the determination of leaf patterns.

If the location of new leaves depends on space, then phyllotaxis is a packing phenomenon that operates not only at the level of the contacts between the leaves after they have formed but also in the determination of the location of the initiation of a new leaf. Genetic and other influences on phyllotaxis would be due to changes in the relative sizes of leaf primordia and the apices on which they are formed. There need not be any genes that are specific to interactions between leaves or are specific to one type of phyllotaxis. The mathematical properties of phyllotaxis (Sinnott, 1960; Erickson, 1983), with which this chapter did not deal, would be a consequence of the packing of leaves, whose bases have different shapes, around apices that have different sizes (Richards, 1951).

The environment and the age of the plant influence phyllotaxis. These influences, or at least some of them, could depend on changes of the relative size of the shoot apices and the leaves that arise on them (McCully and Cutter, 1955; Dale, 1961; Soma and Kuriyama, 1970). This presumably means that the absolute size or volume of tissue that first becomes a leaf is a characteristic of phyllotactic systems. A variety of experimental treatments also modify phyllotaxis, and these, too, could be correlated with changes in the space available for new leaves. As mentioned above, microsurgery changes phyllotaxis when it changes the space available for the new leaves (Snow and Snow, 1931, 1933). Hormones and various substances that have developmental effects have been found to modify phyllotaxis (Meicenheimer, 1981; Mingo-Castel et al., 1984). These effects can be understood as resulting from changes in the growth of the apices: for example, elongation very close to the tip could be enhanced and this would increase the space available for new leaf initiation (Cutter, 1964; Schwabe, 1984). Another problem is that phyllotaxis that is disturbed in various ways often returns to its original state. Some such cases have been accounted for by the spreading of leaf primordia, after they have been initiated, to all the neighboring competent tissues (Snow and Snow, 1962). This spreading tends to stabilize the phyllotactic system by occupying unusual spaces on the apex.

This description of the location of new leaves applies to most plants. Rare exceptions were the subject of special studies by Snow (1952, 1962). He found, for example, that unusual phyllotaxis can be associated with an unusual bending of the apex that influences the space available for the initiation of new leaves. There was also some extension of experimental work to differences of the space required under various conditions (Snow 1952, 1965). However, the parameter by which space was measured was the angle of the apical dome that takes part in leaf formation, while it is possible that a volume or surface area is what is required for the formation of a new leaf. Changes in the initial size of lateral organs could also account for the special patterns formed by reproductive apices (Lyndon and Battey, 1985). The formation of whorls would also depend on individual primordia being small – but whorl formation could also be

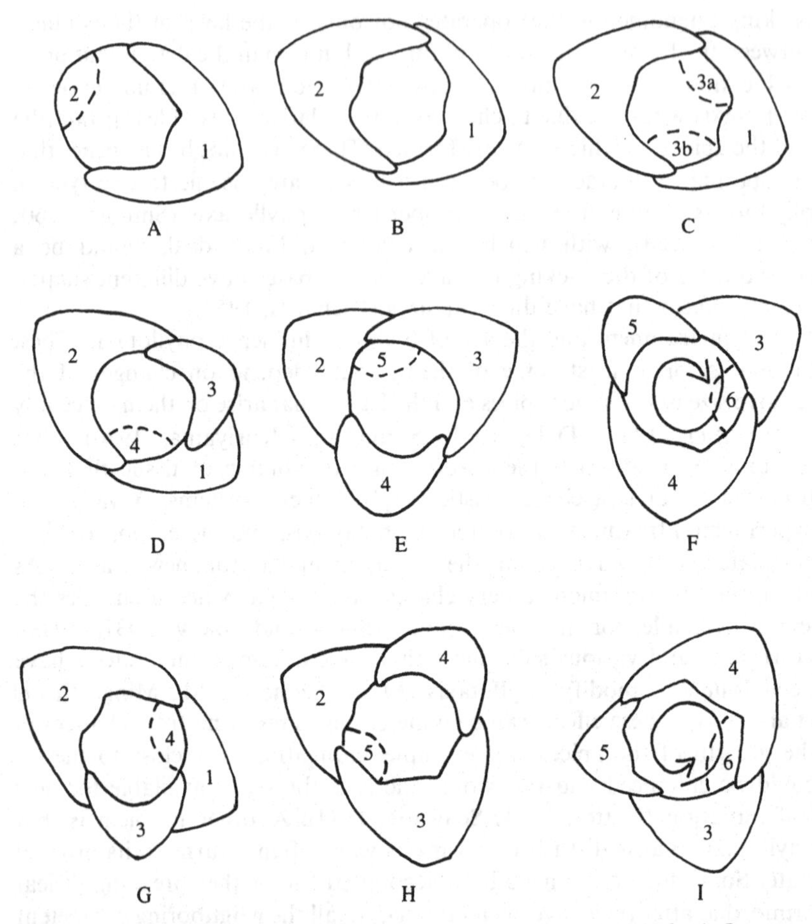

Figure 11.5. The determination of the direction of phyllotactic spirals. A. In a new apex, the first leaf primordium may have a random location. The next leaf (2) is often opposite the first leaf. B. The second leaf spreads around the apex. C. As the apex grows, two equal locations (3a and 3b) become available for the third leaf. These locations are as far as possible from both the second leaf and from the large center of the first leaf. The choice between these two possible locations depends, in many plants, only on chance. D–F. If the lower alternative is chosen the system is self-perpetuating: the next available space is fully determined by the previous leaves and the spiral of new leaves is clockwise. G–I. The same predictability and stability are achieved if the third leaf appears in the top location, but the phyllotactic spiral has the opposite, counterclockwise direction.

viewed as resulting from a single, circular leaf primordia breaking into a number of smaller units after it has been initiated (Rutishauser and Sattler, 1987).

The importance of meristematic space for leaf initiation could also account for the self-perpetuation of phyllotaxis. One indication of this is found in plants in which leaves originate along a spiral (Fig. 11.5). This spiral can have two orientations, clockwise or counterclockwise. On embryonic axes and adventitious shoots this orientation is generally random: it does not depend on heredity and equal numbers of plants of a given species have the two opposite orientations (Beal, 1873; Sinnott, 1960; Rijven, 1968), though there are various exceptions for which no explanation is available (Carr, 1984). Once the spiral starts it is stable. The direction of the spiral in lateral apices is influenced by the pattern on the parent shoot, and this influence often results in an orientation that is opposite that of the parent shoot (Sinnott, 1960). These facts are in agreement with the rule that leaves appear in the first available space, and that the location of this space is determined by the configuration of the previous leaves (Fig. 11.2).

THE ORGANIZATION OF MULTICELLULAR PRIMORDIA

The conclusions reached above put great emphasis on the organization of primordia as multicellular structures. Thus, a final question here is how the cells that form a leaf could cooperate so that a primordium appears only if a sufficient number of cells is available. This is the question of the nature of the organization processes that could require 'space', or a mass of cells. The cooperative activity of many cells suggests that there is an unknown, complex exchange of information between neighboring cells. But as in the case of the organization of promeristems (Chapter 10), the assumption of complex interactions is plausible but not essential. This is illustrated by the following two hypotheses, based on different traits of leaf development.

Physical forces are necessary for the growth of both individual cells and entire primordia, and such forces are also generated by this very same growth (Green, 1962, 1980, 1984, 1987). There have been various suggestions that leaf development is the passive result of physical strains, but these have not been supported by careful observation of the relations between developing leaves and they have been contradicted by the effects of wounds that would be expected to relieve strains (Snow and Snow, 1947, 1951). However, a more detailed view of the structures that restrict the action of physical forces is now possible. Microscopy with polarized light reveals the orientation of the cellulose microfibrils, the structural elements of plant cell walls (Green, 1984). Immunofluorescence studies also show the related distribution of microtubules (Sakaguchi et al., 1988). There are multicellular domains of coordinated microfibril orientation on the surface of shoot apices (Green, 1985, 1986, 1987). Leaf initiation occurs just outside the promeristem wherever domains that differ in their orientation by 90° are in direct contact. The continued initiation of leaves requires repeated reorientation events in the apex, and these reorientations could be generated by continued growth (Green, 1985, 1986, 1987).

This suggests a mechanism that could be the basis of the patterned initiation of new leaves (Green, 1986, 1987): the physical forces generated by growth orient the internal structure of the cells (the microtubules). This determines the orientation of the cellulose microfibrils and they, in turn, act to orient further growth. This feedback could generate a pattern and act as a coordinating influence between neighboring cells (Green, 1985). If this is true, then the differences between apices could be the result of physical traits of the cells and their response to physical forces.

Further work is required to clarify the relation of the reorientation events and the results of the microsurgical treatments that were mentioned above. But the concept of controls by the orientation of cellulose microfibrils already suggests a solution to an unsolved problem of the regulation of phyllotactic patterns. The return of a phyllotaxis disturbed by microsurgery to the decussate state from a spiral state could not be accounted for by known principles of a requirement of leaf initiation for a minimal space: this requirement would predict that once spiral systems are established they should be maintained (Snow, 1942). Green has shown, however, that initiation of leaf primordia on opposite sides of an apex results from orientation events that span the entire apex. The need for space in plants with decussate phyllotaxis is therefore not limited to the size of a single primordium and initiation on one side of the apex could be related to events on the other side.

A second hypothesis concerning the organization of leaf primordia is based on the internal structure of primordia and their determinate growth. Development requires correlative relations with the rest of the plant (Chapters 2, 10). A primordium is known to be a source of auxin, and probably also of other signals, that induce and orient the differentiation of the vascular tissues (Chapters 5, 6). As in the case of promeristems (Chapter 10), this feedback relation between the primordium and the plant could play a major role in the initiation (Lyndon, 1979) and in the organization of the primordium itself. It would mean that a primordium is formed only when the cells collaborate to obtain limiting signals from the plant, even though this collaboration need not require direct interactions between the cells (Fig. 10.4). This hypothesis could also account for the spreading of primordia around an apex (Fig. 11.2): this spreading would consist of the joining of a primordium by competent regions of cells that are not, in themselves, of sufficient size to become organized as independent units. The spreading would occur if the response elicited by a new primordium would suffice to support not only its own growth but would also 'spill over' into neighboring regions.

The two hypotheses concerning the organization of leaf primordia mentioned here are not only compatible with one another – they also do not exclude other possibilities. A characteristic of both hypotheses is that they show that there is no need for a local exchange of complex information between cells, and that in seed plants cells need not be the units of leaf initiation. It seems important to recognize this possibility, even though it does not contradict the idea that local interactions between cells do occur and have important roles.

12

A temporal control of apical differentiation

A few facts can be used to introduce the topic. A shoot apex on a rose bush starts its development by repeatedly producing photosynthetic leaves that resemble one another. Sooner or later the apex changes – it undergoes a process of 'maturation' – and forms the various organs of a flower. Flower formation terminates apical development. This is an example of a differentiation of an apex that results in an ordered spatial distribution of organs along the stem. In a less dramatic form, a similar differentiation is evident in plants in which the flowers are not terminal. For example, in peas the shoot apex of seedling forms leaves with only vegetative buds in their axils; a change is expressed by the appearance of axillary flower primordia rather than vegetative buds. The apex may continue to form leaves and reproductive buds for a long time, so the differentiation need not terminate apical development.

The transition from vegetative to reproductive development is often not a characteristic of an entire plant but rather of individual shoot apices. This must mean that differences between these apices could not be due to the environment but rather to responses to the environment which are dependent on internal differentiation. It is also possible to separate internal processes from the effects of the environment by observing the differentiation of apices in plants kept in uniform conditions. Such experiments reveal an additional point. Though the rose and pea shoots mentioned above undergo a transition from vegetative to reproductive development in any environment in which they are able to develop, this is not true of all plants. There are many herbaceous plants that must grow before they reach a state of 'ripeness to flowering' (Evans, 1960), when they are able to respond to the environment – especially photoperiodic and temperature conditions – by reproductive development (Holdsworth, 1956; Bernier et al., 1981). Thus it is essential to define the change or the maturation of the apices not in relation to the actual formation of reproductive organs but rather by the *capacity* to form these structures when environmental conditions are appropriate. A flowering response to environmental conditions can occur more readily as the plants increase in size (Evans,

1960), indicating that the maturation processes are gradual rather than all-or-none.

The actual changes that occur in these shoot apices are examples of apical differentiation, briefly considered in Chapter 10. But the examples raise a further point: there is a control of apical differentiation that is related to the actual development of the apices, to their age, or to the size of the shoot that these apices have produced. It is this *temporal control of apical differentiation* that is the topic here. As will be seen below, a discussion of apical differentiation as a function of age and size is important for the subjects of this book for two reasons: apical differentiation results in a gradient of traits of tissues along the shoots, and it demonstrates an integrated action of controls from both within and from outside the apex itself.

LEAF NUMBER AS A CRITERION OF MATURATION

Different parameters, such as plant weight, height and age, can be used to measure the size of a shoot at the time the first flower primordia are formed. A parameter that would remain constant under varied conditions could be useful for describing development and it might also be indicative of the controls of apical differentiation. Only one such parameter has been reported: a consistent number of leaves is formed by a shoot apex of a given genotype before it matures and can form flowers (Fig. 12.1; Haupt, 1952; Holdsworth, 1956; Paton, 1978; McDaniel and Hsu, 1976). The most detailed information appears to be available for peas, where the number of leaves before flowering in a given photoperiod is characteristic of different varieties and can be determined by single genes (Reid and Murfet, 1984). Peas grown in different light intensities, including complete darkness, vary greatly in the length of the individual nodes and the height of the plants, and yet the number of the node in which the first microscopic flower primordia are formed remains unchanged. Furthermore, pea seedlings grown in different temperatures develop at different rates, but even this variation has no effect on the node in which the first flower primordia are formed (Milchgrob and Sachs, unpublished). These observations suggest that there is a close relation between the apical maturation expressed by reproductive and the formation of leaves by these very same apices.

It was mentioned above that maturation is a process that occurs in individual apices, but it could still be influenced or determined by characteristics of the rest of the plant. It is relatively easy to check, however, which leaves are correlated with the maturation of apices. Pea plants were cut and buds removed so that laterals developed even though there were stems and leaves morphologically above them (Milchgrob and

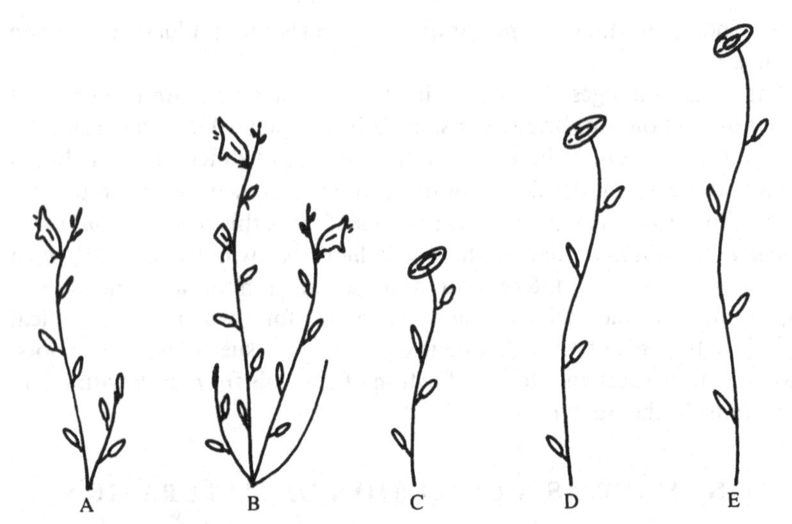

Figure 12.1. The correlation between the location of flowers and parameters of plant size. A, B. Different stages in the development of a plant that branches close to the roots. Flowers appear on the branches that have reached a characteristic size, with no reference to the size of the plant as a whole. C–E. Three plants of the same genetic constitution which grew in different light intensities. The appearance of the first flowers was not clearly related to the height of the plants. The best correlation was with number of nodes, and thus the number of leaves, which separated the growing shoot apex from the roots.

Sachs, unpublished). These treatments resulted in breaking the normal correlation between overall plant size and the number of nodes that separate an individual developing apex from the root system. The results of such treatments were clear cut: the location of flower development was correlated only with the number of leaves (and nodes) that separate an apex from the roots. The presence of large mature leaves that were not on the direct line to the roots had no influence on the location of the first flowers. This result is in accordance with observations of many untreated plants of various species: lower shoots do not form flowers even when they are parts of large plants (Fig. 12.1A). It thus appears that whatever the nature of the mechanisms that determine apical differentiation, they do not depend on correlative effects of the entire shoot system. Critical events must occur in the apex itself or depend on some aspect of the distance of this apex from the roots.

Leaf numbers on lateral branches

A relation between leaf number and reproductive development is also suggested by the location of the first flowers on lateral buds that have been allowed to grow as dominant branches (Fig. 12.2A). In most plants it is possible to cause the development of specific laterals by removing the shoot above them (A release from apical dominance, Chapter 2). In pepper (Ryleski and Halevy, 1972) and tobacco (McDaniel and Hsu, 1976; McDaniel, 1978) there is a remarkably constant total number of nodes separating a flowering apex from the root system, even when the apex is that of a lateral branch. Thus, the higher the location of a lateral on the plant the fewer the leaves it forms before flowering. In other words, in these plants the number of the leaves on the lateral plus the number below its branching point equals the number of leaves formed by an undisturbed main shoot apex.

There are many exceptions in which lateral branches, even when they become dominant shoots, do not behave as mere replacements of the main apex. Obvious examples are plants in which flowers terminate the development of the main shoot apices. In these, continued development depends on the growth of lateral branches (Fig. 12.2B) which often appear right below the terminal flower – and yet they form one or more nodes, carrying photosynthetic leaves, before they terminate in flowers and the cycle of 'sympodial development' is repeated.It follows that there must be some way in which lateral buds high up on the plant can be 'rejuvenated' so their maturation processes start from a new level (Nozeran et al., 1971).

The opposite behavior of laterals, a maturation before an 'appropriate' nodal distance from the roots is achieved, is also common. In untreated plants, laterals may form flower buds after fewer leaves than would be formed on the main shoot apex above them. This is often due to environmental signals determining the expression of the capacity for reproductive development independently of the precise developmental state of the various apices. But not always: the lateral buds of pea seedlings can be released from dominance, in uniform laboratory conditions, by the removal of the main apex and such buds form flower primordia long before they are separated from the roots by the number of nodes typical of the variety (Fig. 12.2C). It is as though the branching node itself has the same effect as a number of leaves, but this number varies depending on the location.

Even though not all lateral shoots behave in the same way as the main apices, they too show that leaf initiation is correlated with apical maturation. This could mean that leaf primordia induce changes in their parent apices. It is also possible that leaves are only a measure or expression of apical events, such as growth and cell division, and they themselves have no effects on apical differentiation. These possibilities call for simple experiments in which leaves are removed and the maturation of apices is observed. In peas and tobacco, removing macroscopic leaves has no effect on the location of the first flowers (Sprent, 1966; McDaniel, 1980). Removing leaf primordia that are barely visible, however, does delay the flowering node: in many of the treated plants an additional leaf

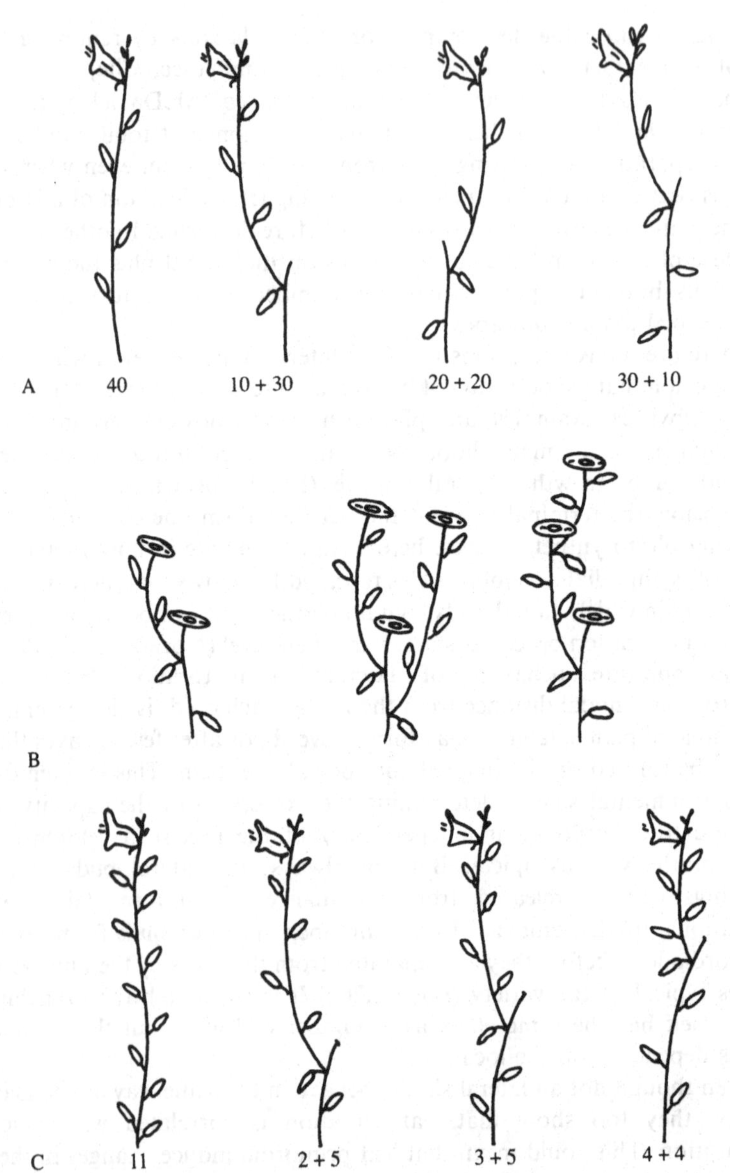

Figure 12.2. Transitions to flowering in lateral branches of various plants. The numbers below the plants refer to the nodes formed before the appearance of the first flower. A. lateral branches released from dominance in a plant such as tobacco. The intact plant blooms when 40 node separate the apex from the roots. Lateral branches grew when the plants were cut; these branches bloomed when the total number of nodes remaining on the main stem, plus the ones on the growing lateral branch, was approximately 40. B. In many plants flowering terminates apical growth; continued development depends on lateral branches. These laterals produce additional leaves before they flower. The number of leaves varies: it may

is formed, 'replacing' the removed leaf, before flower primordia appear (Milchgrob and Sachs, unpublished). Leaves are only 'replaced' if they are removed at a very early stage, at a time they are barely visible. But damage to the apical dome which does not remove leaf primordia also delays flowering, often requiring an additional leaf before flowering is achieved. Thus, the results available at present do not support any assumption of a direct effect of developing leaves on the maturation of apices, but further work is clearly called for.

EXPERIMENTAL MODIFICATIONS OF SIZE

Rooting apical cuttings is a simple way of changing the distance from a shoot apex to the roots. Such rooting delays flowering, even in the appropriate environmental conditions (Robinson and Wareing, 1969; Schwabe and Al-Doori, 1973; Miginiac, 1978). In tobacco, where quantitative work has been carried out (McDaniel and Hsu, 1976; McDaniel, 1978), short cuttings form as many leaves before flowering as do seedlings. On the other hand, other cuttings taken from higher up on the stem, close to the flowers, only complete the same number of leaves as they would on the intact plant – they are not affected by the new roots (Fig. 12.3A). These results indicate a determination which occurs a few nodes before the flowers become apparent. In other plants and conditions, however, the results of comparable treatments are less clear cut. It is generally true that the formation of new roots delays flowering, but the effects vary and do not correspond quantitatively to the events on the intact plant (Fig. 12.3B; Robinson and Wareing, 1969). Rooted cuttings of grape vines remain mature and do not go through a juvenile phase (Mullins et al., 1979). Another example is a rooted inflorescence of *Coleus* (Sachs, unpublished): the lateral buds just below the inflorescence grow out as 'rejuvenated' shoots and form photosynthetic leaves before they, too, terminate in an inflorescence (Fig. 12.3C). This shows that apical changes can be reversed, not only continued. The number of leaves formed by the 'rejuvenated' buds, about four or five pairs, varies depending on conditions, but this number can be smaller than that found between the roots and the inflorescence of the parent plant.

Apices can also be cut and grafted in various new locations relative to the roots (Fig. 12.4; Haupt, 1954; Paton and Barber, 1955). Grafts that move apices closer to the roots tend to delay flowering and grafts that

be small in poor conditions (on the left) and considerably larger in the very same plants when they are vigorous (on the right). C. In peas the behavior of the lateral branches is fairly regular, but unlike tobacco they form fewer leaves before flowering than would have been formed by the main apex. These lateral branches behave as though the branching node itself is the equivalent of a few nodes along an intact shoot. No simple rules of additive distances apply.

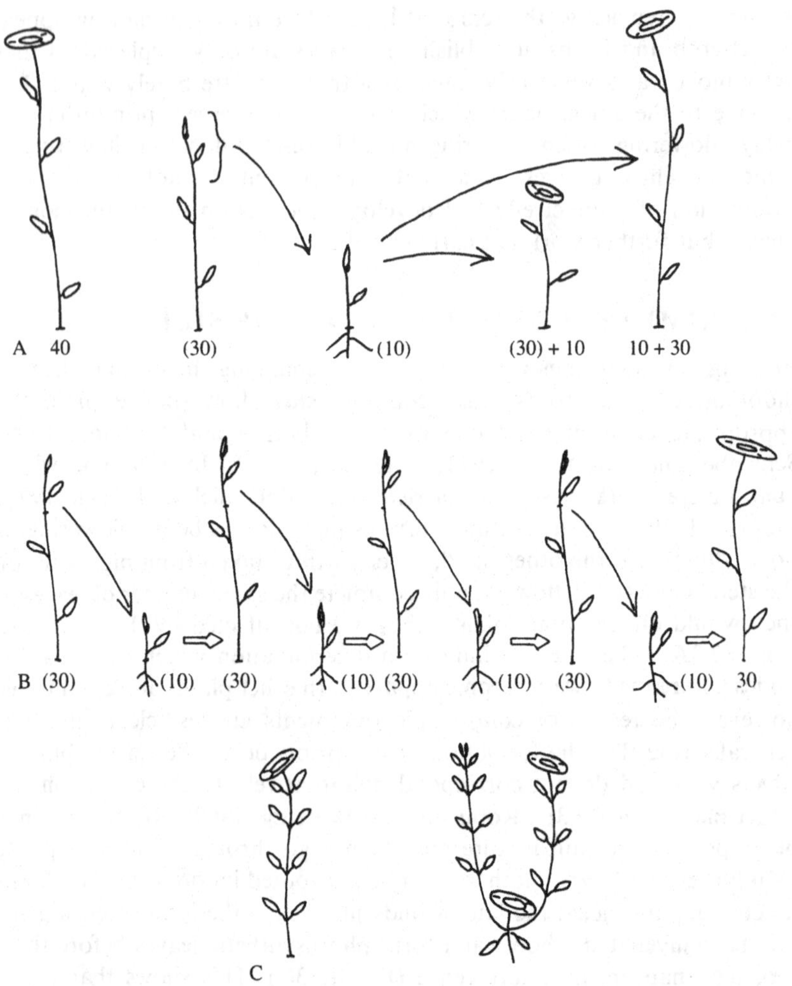

Figure 12.3. Rooting of cuttings as a means of modifyng the location of apices. The numbers below the plants refer to number of nodes; in parenthesis are the numbers of nodes immediately before the plants were rooted and other numbers refer to the state of the plant when it first bloomed. A. The tobacco variety which was used blooms after it has formed 40 nodes. Two types of behavior occurred in rooted cuttings taken from a plant that had not yet started to flower: the cuttings either formed the complete number of nodes characteristic of their original location on the plant, or, in the very same experiment, the cuttings 'forgot' their past and flowered in accordance with the number of nodes on the new rooted plant (based on McDaniel). B. Repeated rooting of cuttings was used for repeated delays of the flowering of the main apex. At least in *Ribes*, such plants eventually flowered with fewer nodes than would be present on the intact controls (based on Robinson and Wareing). C. When the inflorescence of some plants (such as *Coleus*) was rooted it formed lateral branches which would not have grown on the intact plant (left) and did not flower for some time. Thus, rooting of cuttings can not only delay but can also reverse the transition to the reproductive state.

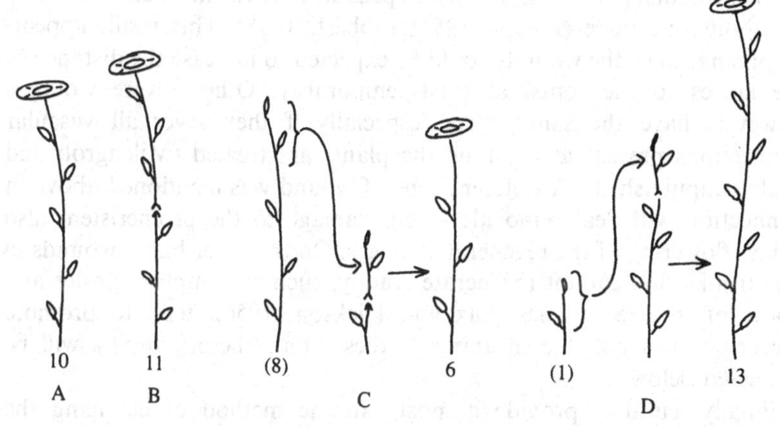

Figure 12.4. The use of grafting to modify the location of shoot apices. The numbers below the plants refer to number of nodes; in parenthesis are the numbers of nodes that separated the shoot apex from the roots before it was cut and other numbers refer to the state of the plant when it first bloomed. A. Control; an intact plant that blooms when it has developed 10 nodes. B. Cutting and regrafting of the upper part of a shoot in its original location has no obvious effect on the relation of the roots. Yet such treatments do postpone flowering by about one node. C. The upper part of a shoot of a large plant was grafted onto a young plant. Flowering was often at an intermediate plant size – larger than that expected if the apex had not been disturbed, but smaller than that characteristic of the new plant and the age of the roots. D. A shoot apex of a seedling was grafted on to a large plant. Flowering was influenced by both members of the graft: the shoot apex formed flowers later than expected of the mature plant but earlier than would be required for the apex of the seedling to form its full complement of leaves. Based on various experiments, especially Haupt (1954).

increase the distance to the roots have the opposite effect. These results support the suggestion that apical distance from the roots is an important control of the transition from vegetative to reproductive development. However, with the exception of work on tobacco, the results do not support a simple quantitative relation between flowering and the distance between apices and roots at any given time. Instead, there are indications that the apices are influenced by their past determination, before they were grafted in their new location, and that this determination can be expressed by quantitative modifications of the location of flowers. For example, apices taken from seedlings and grafted on mature trees still do not bloom for some time, but they do bloom sooner than apices that are grafted on small seedlings: *both the new location and past developmental history influence the location of the first flowers.*

Wounding stems is a third method that could alter the physiological distance of apices to the roots. The most extreme wounds are cuts that separate the plant into two parts, followed by a regrafting of the tissues in their original place (Fig. 12.4B). In peas such treatments delay flowering by about one node (Haupt, 1954; Libbert, 1955). This result appears surprising, since the wounds could be expected to increase the distance of the apices to the roots, at least temporarily. Other severe wounds, however, have the same effect, especially if they sever all vascular connections present at the time the plants are treated (Milchgrob and Sachs, unpublished). A different type of wound was mentioned above in connection with leaf removal: severe damage to the promeristem also delays flowering of the regenerated apices. On the other hand, wounds in tree trunks that cannot regenerate readily, such as complete girdles and grafts of inverted tissues (Sax and Dickson, 1956), tend to promote flowering. The possible meaning of these contradictory results will be discussed below.

Finally, cultures provide a most extreme method of changing the relations between a tissue and the rest of the plant. Tissues from various parts of a plant can regenerate apices and these can be scored for their flowering behavior. Here, again, much of the information comes from tobacco, where adventitious apices can be induced readily. Tissues close to the inflorescence formed flowers directly, with no intervening leaves, while tissues from the base formed vegetative apices (Chouard and Aghion, 1961; Tran Thanh Van, 1973). The intermediate tissues formed different numbers of leaves before flowering, the number decreasing as the original distance from the roots increased, though such gradients of flowering are not always obvious (Wardell and Skoog, 1969). These results suggest or confirm two important conclusions. The first is that the transition to flowering can be a stable or determined trait (Wareing, 1978; Meins, 1986) that is maintained even in mature cells and even when they have been subjected to culture conditions (Mullins et al., 1979). Evidence supporting this conclusion is also available from plants that do not regenerate apices readily. In *Hedera* callus grown from shoots that do and do not carry reproductive organs differed in various characteristics, showing traits of the original tissues that are maintained in culture (Chapter 4; Stoutemyer and Britt, 1965; Banks, 1979). The second conclusion from tissue culture work is that the apical transitions can be gradual, a potential for flowering increasing as one proceeds up the stem. Additional expressions of this gradual change will be mentioned below.

THE POSSIBLE ROLE OF THE INTEGRATED DISTANCE TO THE ROOTS

The results summarized above indicate that the transition to reproductive development is influenced by two factors internal to the plant: a physiological distance of the apices from the roots, for which the number of nodes is a rough measure, and a determinate or stable change within the

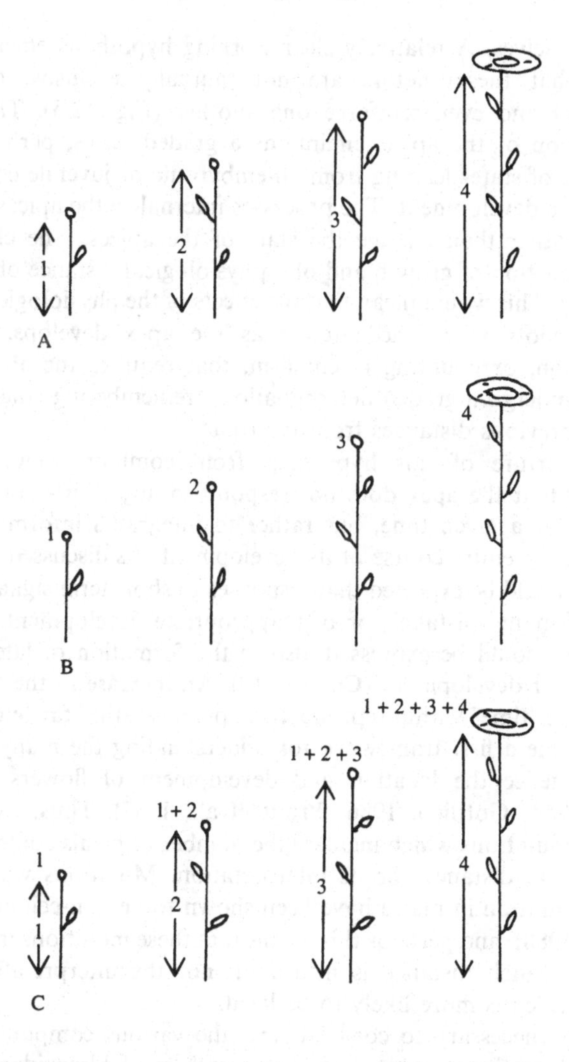

Figure 12.5. Hypotheses concerning the internal control of the location of the first flowers. A. Control by the physiological distance of the shoot apices to the roots. This distance increases as the plant develops; when it reaches the threshold of '4 nodes' the apex changes to the reproductive state. B. Control by an internal counting mechanism in the apex. Cell divisions or some other characteristics of apical development are counted. When the four nodes have been produced the apex reaches a threshold of '4' and changes to the reproductive state. C. Integrated control by an apical 'memory' of past distances to the root. The apex changes gradually, as a function of all its past distances to the roots (the function presented is a simple sum, but this is intended only for clarity). Thus apical changes are a function of both the correlative effect of the distance from the roots and an internal counting mechanism in the apex itself. It is only this 'integrated' hypothesis that is consistent with available evidence.

apices themselves. A relatively clear working hypothesis emerges if it is assumed that these factors are not mutually exclusive but rather complement and even reinforce one another (Fig. 12.5). The internal determination of the apices maintains a graded series, perhaps even a continuum, of states leading from an embyronic or juvenile condition to reproductive development. The processes internal to the apices, however, maintain rather than change the state of the apices. The changes are always a function of growth and of a physiological distance of the apices to the roots. This would mean that the effects of the physiological distance from the roots would accumulate as the apex develops. It is this accumulation, or counting mechanism, that requires the ability of the apices to undergo a graded determination, 'remembering' their previous states, or previous distances from the roots.

The departure of this hypothesis from common views is in the suggestion that the apex does not respond to any short-term signal, to conditions at a given time, but rather to integrated information of its states over the entire course of its development. As discussed in the final chapter, it could be expected that responses to short term signals could be sources of many mistakes, or of inappropriate development. The same mechanisms could be expressed also in the formation of lateral organs during flower development (Chapter 10). An increase in the 'integral of distance over time' within reproductive apices is rather far-fetched, but it could provide a first framework for understanding the many mutations which influence the location and development of flowers (Reid and Murfet, 1984; Gottlieb, 1986; Pruitt et al., 1987). Thus, the common homeotic mutations which increase the number of petals could act on the 'physiological distance' or its interpretation. Mutations which change apical maturation in maize have been shown to be non-cell-autonomous (Poethig, 1988), and perhaps this means that these mutations influence the signals by which distance is measured, not the interpretation of this distance, which is more likely to be local.

It is now necessary to consider how the various components of this 'integral' hypothesis are in accordance with available evidence. Results mentioned above, especially concerning apices grafted in new locations and tissue culture work, prove that apices, and even the mature tissues that they form, can have determinate or stable information concerning the transitions they have undergone towards reproductive development. This evidence, however, has been taken to indicate a progression of changes within the apex (Robinson and Wareing, 1969), and while this is an attractive hypothesis it is not required by any known facts. One concrete suggestion concerning the cause of the internal changes was that cell divisions in the apex are counted in some way, until a reproductive state is reached. However, it is possible to increase cell divisions in the apical dome by wounds, and, as mentioned above, such wounds delay rather than enhance apical maturation.

The hypothesis requires that the change in the apices is gradual and that small changes should be 'remembered' or stored by the apices. This is supported by direct evidence of apices regenerated in culture, by the behavior of apices grafted in various locations and, as will be mentioned below, by apical differentiation expressed by phenomena other than reproductive development. However, there are also results that indicate threshold transitions between apices which are and which are not committed to flowering (McDaniel, 1978, 1984). These results can be readily accounted for when it is realized that determination itself is a quantitative term, a function of the method used to disrupt the system. It is quite possible, therefore, that in tobacco apical determination is gradual, but rooting disrupts all its expressions below a certain value. Perhaps an apparent threshold of stability appears when apical structure, and not only its physiological state, has started to change towards reproductive development. Thus, evidence of threshold determination does not contradict other evidence for gradual transitions in the differentiated states of shoot apices.

The measurement of physiological distance

The diverse facts concerning the 'physiological distance from the roots' that signals apical change are in accordance with the possibility that distance is a function of the contacts between vascular channels, presumably sieve tubes, through which materials are transported. This would account for distance not being related to stem length in any simple way. The relation to the number of leaves could be understood, as could the fact that the correspondence between apical differentiation and leaf number does not hold under all conditions. Leaf number reflects the number of nodes, and it is at the nodes that most ends of vascular channels occur. Contacts between channel endings might be most common when a lateral bud forms contacts with the existing stem channels, the channels that led to a removed shoot. This could be the reason why a branching point is, in peas, the equivalent of a number of leaves or nodes. Finally, the importance of vascular contacts could account for the 'rejuvenation' which occurs when stems are wounded and when new lateral growth occurs after the main apex has terminated development. In these cases, an entirely new vascular system is formed from the cambium (Benayoun et al., 1975; Sachs, 1981a) and this might be physiologically shorter than any previous system that developed and formed contacts as the plant developed. The hypothesis that contacts between sieve tubes are a major aspect of the physiological distance of the developing apices from the roots could thus account for available data, but it is hardly proven by any critical experiment. It does, however, suggest anatomical observations of the connections between sieve tubes along the stem.

Even if development were to be related to vascular contacts, there is need to consider how the distance to either the base or the tips of the roots could be measured. There are various ways by which this measurement could occur, but there is no clear evidence to support or reject a number of reasonable suggestions (Khait, 1986). For example, a balance of two (or more) hormones that originate in the roots could be a function of distance. If one hormone were impeded more than another, their balance in the shoot apices would change gradually as development increased the distance from the roots. Possible hormones that originate in the roots are gibberellins and cytokinins (Chapter 3), and there is evidence that the addition of these substances to plants changes the reproductive behavior of apices (Bernier et al., 1981). Gibberellins are also known to influence other developmental changes, considered below, associated with the distance of apices from the roots, especially the juvenility of *Hedera* (Robbins, 1960; Frydman and Wareing, 1973; Rogler and Hackett, 1975). The effects of these substances vary a great deal, however, and no general conclusion about a balance that determines flowering is warranted at present. Distance could also be measured by the relation between two oscillations of different wavelengths (Goodwin and Cohen, 1969). Such oscillations could be pulses of substances such as hormones, but they could also depend on electrical or other signals. The possibility that distance depends on oscillations is certainly attractive, and it could account for phenomena other than the control of apical differentiation (Chapter 13), but it is yet to be supported by concrete evidence.

OTHER EXPRESSIONS OF MATURATION

It is now necessary to ask whether there are other developmental expressions of the increase of the distance of shoot apices from the roots. Such additional expressions could, of course, be of interest in themselves. They could also be of direct use as evidence concerning the working hypothesis outlined above: they could add essential information, especially about a gradual differentiation of apices that is not expressed by the all-or-none phenomenon of flowering.

The maturation of woody plants

A most obvious temporal change of developing apices is the transition from juvenile to mature development (Goebel, 1900; Wareing, 1959, 1978; Moorby and Wareing, 1963; Zimmerman, 1972; Thomas and

Figure 12.6. Juvenility and maturation in woody plants. A–C. *Hedera helix*. The plant that develops from the seed (A) is a climber that is quite different from later growth in the form of a bush (B), the only form which carries flowers. All possible transitions, expressed by the shape of the leaves (C) can be found on the very same plants. D, E. *Pinus canariensis*. Leaves are attached directly to the main stems in juvenile branches while in the mature form they appear only on specialized side branches (brachyblasts). Buds that grow at the base of mature trees are juvenile (D). Higher up on the trunk all bud growth has the mature form. There is no clear line separating the two developmental types: in the transition region both juvenile and mature buds may develop side by side.

Vince-Prue, 1984). This transition is common to woody plants and is generally reversed only when embryos are formed. It may be best illustrated by the well-known example of English ivy (*Hedera helix*) which has two growth habits formed on the same plant and formed even by the same apex (Fig. 12.6A–C; Goebel, 1900; Sinnot, 1960). Germinating seeds always develop as climbing plants. They have heart-shaped leaves arranged in two rows on flattened stems which attach to rocks and tree trunks by short, sticky roots; there are never any flowers on these climbing shoots. The second growth habit is a bush that develops from the early climber phase. This bush has ovate or elongated leaves arranged in a spiral on stems that are round in cross-section and are stiff enough to carry themselves at any angle. The shoots of the bush initiate roots only with great difficulty and they form flowers every year. *Hedera* thus has two developmental phases, a juvenile one that develops from the germinating seeds and a mature phase which is distinguished by reproductive development, but also by various other traits. These terms, 'juvenile' and 'mature', are somewhat confusing: it is the juvenile form that develops first, so its tissues are older than those of the mature phase. The apical development of plants means that young tissues are found on mature plants and the oldest, in some senses most mature, tissues are the juvenile ones.

Hedera is an outstanding example of a transition from juvenility to maturity that is common to many if not most woody plants (Goebel, 1900; Wareing, 1959, 1978; Zimmerman, 1972; Thomas and Vince-Prue, 1984). Observations and experimental work have dealt primarily with the location of the first flowers on the plant, and the term juvenility is often defined in relation to the absence of reproductive development. However, a capacity for reproductive development is only one of many expressions of shoot maturation. Thus, the leaves of tree seedlings are often different from the leaves of the mature trees that develop from these very same seedlings. This stands out in pines (Fig. 12.6D,E), where the juvenile leaves may have a striking silver color and they differ from the needles of the mature trees in being carried directly on the stems and not on short branches (brachyblasts). In deciduous oaks the leaves of the juvenile tree may lack an abscission zone (Schaffalitsky de Muckadell, 1954). Most trees undergo additional changes when they are many years old. In these trees there are no dominant, leading branches and the increase in height of the plant ceases.

The various expressions of transitions from juvenile to mature phases in woody plants could be additional manifestations of the same internal controls considered above, the controls being expressed by the location of flowers. This is only an hypothesis, but it could offer experimental systems for the study of apical differentiation. It is thus important that changes associated with maturation are characteristic of apices, not of whole plants, and that these changes depend on the distance of these apices from the roots. The traits associated with tree maturation are stable or determinate at the tissue or apex level: buds growing from the base of cut trees have the juvenile traits characteristic of the time the tissues were formed (Fig. 12.6D), including the absence of reproductive structures. This is true even of adventitious buds, ones formed from tissues that were not meristematic, and in this sense were mature, for a very long period. Finally, the changes from juvenility to maturity are often gradual: this is expressed by the form of the leaves (Fig. 12.6C), the ease with which cuttings initiate adventitious roots, and the gradual decrease in the relative dominance of leader shoots. Gradual changes are also indicated by there being more than one maturation event in the development of many trees: in pines the seedlings are distinguished by juvenile characters, while the capacity to form reproductive organs appears only later. Finally, the shape of the tree changes only after many years of reproductive development.

Juvenility and maturity in herbaceous plants

Comparable maturation changes can be found in herbaceous plants, though the changes are necessarily less gradual or detailed. Thus, there are often regions of bracts, or even specialized photosynthetic leaves, that precede the formation of flowers and are part of an inflorescence only when the term is used in a broad sense. Even in herbaceous plants it is often easier to root young shoots that do not, and cannot, carry flowers. Available experimental evidence concerning changes of the location of apices relative to the roots, however, appears to be limited to the formation of flowers, and these were considered above.

An additional gradual change which occurs as many plants mature concerns the shape of the leaves (Goebel, 1900), a phenomenon known as heteroblasty. The changes involve greater lobing or, in compound leaves, a larger number of leaflets. These leaf changes are influenced by the supply of sugars or by the vigor of the plant as a whole (Goebel, 1900; Allsopp, 1964). Heteroblasty, furthermore, does not display the determination characteristic of traits related to apical juvenility and maturation: new leaves following rooting or damage are often typical of the new state of the apex, with no clear effects of its developmental past. It is therefore likely that leaf shape differences on a given plant can be the function of at least two different control systems: the change from juvenility to maturity considered here, and the relative vigor which influences primordial leaf development directly. Heteroblasty is thus not directly related to the apical maturation considered in this chapter.

Sympodial development of shoot apices

The common phenomenon of sympodial development may depend on the same general control. Here apices cease development after a certain number of leaves are formed, and these apices are replaced by lateral branches. The new apices form a number of nodes and the process of replacement is repeated. The apical change is often associated with the formation of a flower or inflorescence (Fig. 12.2B), but it may also involve the abortion or cessation of apical activity without the formation of any special structure. This determination or apical change does not occur close to the root system and in this and other ways it resembles expressions of apical maturation. The special characteristic of sympodial growth, however, is that the lateral apices are able to continue growing – are 'rejuvenated' – in some way that is not strictly dependent on their location on the plant. The new buds have new, continuous vascular channels, formed from the cambium, and perhaps this 'rejuvenation' is another indication of the role of vascular contacts in apical differentiation.

13

Generalizations about tissue patterning

CONTROLS OF CELLULAR DIFFERENTIATION

Tissue patterning is an orderly, controlled expression of various potentials for cellular growth and differentiation. It is often assumed, or at least implied, that the capacity of plant cells to differentiate is unlimited and that each event is specified by the precise conditions the cells are in at some critical, early period of their development. This dogma is certainly not borne out by the discussions above, and an alternative set of generalizations is required. Thus, before the patterning processes that were arrived at in the previous chapters are summarized and discussed it is necessary to consider what developmental alternatives are available to plant cells and what determines their expression.

Totipotentiatility and limited developmental alternatives

Organized development, like other major functions of a plant, is limited to special tissues. The cells of these developing tissues are of distinct, differentiated types: they have unique characteristics and undergo only limited developmental changes, even in experimental conditions (Chapter 10). Furthermore, it is not just development but also its various component processes that are performed by cells of different specialization. Organogenesis occurs only in promeristematic, or embryonic, regions – in small cells with a dense cytoplasm, no obvious polarity, and slow growth. Most increase in volume or growth occurs in primary meristems where development is rapid, polarized and generally of limited duration. Cambia, whose cells differ from those of promeristems, are responsible for increase in diameter. Only growth and cell division that do not result in organized structures can occur even in cells not specialized for development (Chapter 4).

Transitions between the various differentiated states of plant cells are always limited, and most transitions never occur. This applies even to embryonic or promeristematic cells: they can be the origin of all tissues, but only indirectly, through the slow formation of primary meristems

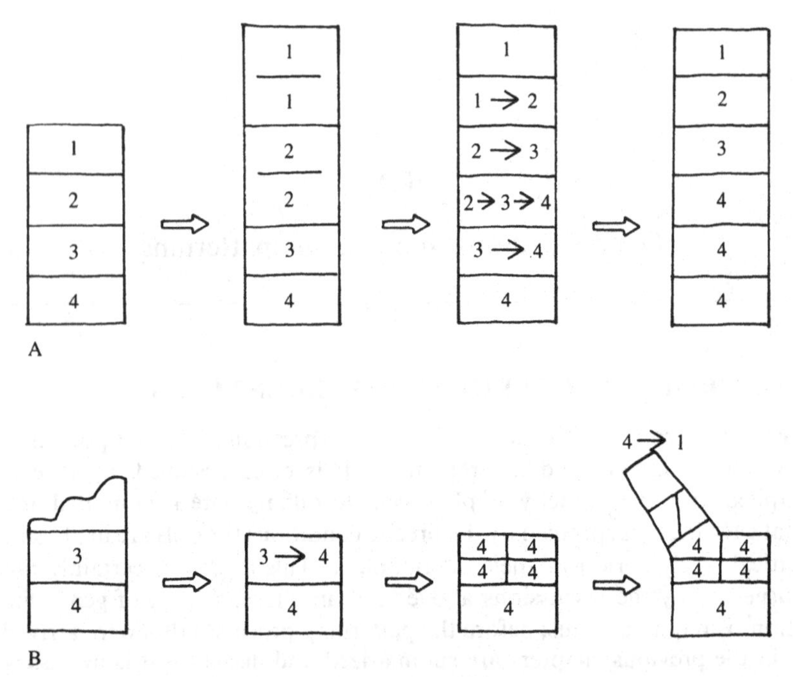

Figure 13.1. Totipotentiality is consistent with the limited changes of differentiation available to plant cells. A. Simplified 'plant'; the numbers represent various differentiation states, '1' being the embryonic or apical. The addition of cells during development changes the relative positions of all cells. New locations lead to limited changes of differentiation: cells maintain their state or change to a state represented by the next higher number. Cells skipping over stages and reversals of differentiation are not common. B. Rules concerning changes of differentiated states are maintained even in a 'plant' that has been severely damaged. yet there is one important exception: cells can change from one mature state, '4', to the embryonic state from which all other cells types arise, albeit indirectly. It is this possibility that makes plant cells totipotent even though transitions between differentiated states are always limited.

(Fig. 13.1). It is hardly surprising that the repertoire of possible changes of mature cells, ones not normally involved in development, is no less limited. There are conditions in which some mature cells divide to form promeristematic, embryonic cells – and it is through this transition that the indirect formation of entire new plants is possible. Plant cells can thus be totipotential, but this totipotentiality does not mean an ability to change directly into all the various differentiated cell types found in a plant.

It is likely that differences of gene expression are primary events of cell differentiation. This would mean that determinate, mature cellular states depend on stable configurations of gene expression. However, these statements need not imply that cell differentiation always depends on gene expression and its stability. Important, early differences between cells are the orientation and degree of polarization (Chapters 5, 6). Since gene expression has no known orientation, polarity differences between cells are due to cell organelles or to local membrane structures. Furthermore, there is evidence that the determination or stability of the polar state is a function of positive feedback relations that need not involve gene expression (Chapter 5). Finally, differences in cell size, form, and even function, need not be end points of differentiation processes or of stable modes of gene expression. Instead, these differences can result from processes that have been left 'incomplete' to various degrees. For example, the parenchyma of the vascular system appears to have undergone an induction to differentiate as vascular channels, a differentiation which was completed only in a small part of the tissue (Chapter 6).

These views of plant cell differentiation are hardly surprising, but they are in sharp contrast with concepts stated or implied in many publications. It is assumed that there are undifferentiated states from which all differentiated cells can be readily derived. Tumors, for example, are said to consist of such 'undifferentiated cells'. The term 'undifferentiated' is used without explanation, let alone definition. It would perhaps be appropriate to call cells undifferentiated if they lack any stable characteristics and are able to undergo all differentiation processes readily – but there are no such cells in plants. It is thus best to refer to all plant cell types as differentiated from one another, the zygote and other embryonic cells being no exception. For the following discussion of patterning in plants an important point is that very few transitions between differentiated states are available to any given cell.

Specification of differentiation processes

Next to be considered are the factors that determine or control the transitions between differentiated states. The term 'controls' was used above in an intuitive way and some explanation or broad definition is required. Since there are different alternatives open to any developmental system, there must be conditions, or controls that determine which alternative is actually expressed. It is expected that the same differentiation may prevail for various reasons, so controls should not be characterized by the specific differentiation they induce. A characteristic of controls that play a role in organization must be the coordination of events, along either spatial or temporal axes.

Controls that coordinate developmental events may be exerted by essential conditions, such as metabolites, or by special developmental signals. These two possibilities certainly overlap, but a convenient definition would be that a requirement for a signal, unlike a requirement for an essential metabolite, could be circumvented by an appropriate mutation. Since many developmental processes require the same metabolites, the very fact that signals are not necessarily essential means that they can have specific effects on development. Controls by signals are therefore expected, though this does not rule out some roles for metabolites.

Controls can be readily divided into two groups. The first is intracellular conditions or processes (Chapter 1): since all cells are derived from an original cell, these intracellular controls must depend on or reflect the past developmental history of the cell. The second group of controls are intercellular: they include all the signals that a cell receives from the surrounding tissues. A third possible group of developmental controls would be environmental conditions, but as explained in the Introduction, the environment regulates the development of the plant as a whole, and could not detrmine the patterning of individual cells and tissues, the subject considered here. The relative role of the intra- and inter-cellular controls can now be discussed separately; but the purpose of the separation is only to form a basis for the synthesis of a general picture below. The possible contributions of intracellular controls mentioned in Chapter 1 should be considered in reference to known facts. The state the cells have reached in any given time can become determined, it can be 'remembered', even when conditions change. There is good evidence for such determination in plants, especially as expressed by callus growth (Chapter 4), polarity (Chapters 5, 6) and the development of shoot apices (Chapter 12). Furthermore, this determination can be quantitative: there are small differences between determined states and the influence of conditions can accumulate (Chapters 6, 12). Intracellular controls can also be expressed by the competence of the cells to respond in various ways to spatial signals of other tissues and to environmental conditions. Finally, intracellular controls could be expressed by complex developmental programs that are executed regardless of the surrounding tissues (Chapter 8). Yet it is remarkable how few developmental programs have been found, and how limited are their roles.

Turning to intercellular controls, or spatial correlations, there is no lack of indications and even of concrete evidence that the developmental fate of a cell depends on signals of neighboring cells, tissues and organs. Thus, root and shoot apices induce the initiation and promeristematic development of one another (Chapters 2, 3). Developing leaves and root apices induce the polarization of the axes that connect them to the plant and induce the differentiation of some of the cells of these axes as vascular tissues (Chapters 2, 5). In these cases the nature of some of the signals involved is known – and they are substances, hormones, which act at low concentrations (Chapter 3). These spatial effects are expected; what is unexpected is the low specificity of the known spatial effects on differentiation. Developing organs induce the differentiation of subtending tissues, but this induction is of entire developmental processes not of their components, such as cell divisions, cell elongation, etc. Furthermore, the induction is of entire tissues, not individual cell types, and the inductive effects are not typical of any narrow, definable stage in the development of the organ: where there are differences in the inductive effects they are only quantitative. There is evidence or even proof of the inductive effects of known hormones – but these hormones elicit many different cellular responses rather than one specific developmental process or the differentiation of one type of cell.

The integration of intercellular and intracellular controls

All this is a far cry from the common dogma that developmental events are evoked by precise balances of specific substances (Went, 1938). It is remarkable how little evidence can be found for any part of this dogma. The best known support comes from controls of the formation of shoot and root apices in culture, found to be influenced by the balance between auxins and cytokinins (Skoog and Miller, 1957). Even this is not a balance of substances specific only to these differentiation processes. Other reported cases are at best of limited applicability. Substances certainly do influence differentiation, but the effect is not specific either to the substance or to its precise concentration.

The discrepancy between the available evidence and the common dogma could be due to present ignorance. It is likely that this is at least part of the answer: available information is certainly limited and some examples of unsolved problems will be mentioned below. But the discrepancy means that the dogma itself should be questioned. A precise configuration of signals would be required if it were the only specification of cell fate. Additional specification could depend on intracellular rather than intercellular processes. Although it was concluded above that detailed intracellular programs have only a limited role in the specification of differentiation, a detailed specification could be the product of an integrated action of spatial signals and the quantitative intracellular determination or memory (Fig. 13.2; Sachs, 1988c). This would occur if the influence of spatial signals were gradual and its effects would accumulate during development. And it is precisely such gradual specification that was found in developmental systems considered in the chapters above.

Such 'integrated controls of development' could have the necessary information content needed to specify the fate of many cell types with few, unspecific signals. One reason is that an entire dimension would be added by cells that undergo an induction of differentiation for different lengths of time. Furthermore, it would not be just concentration at any given time that would matter. Instead the cell could respond to changes with time, to

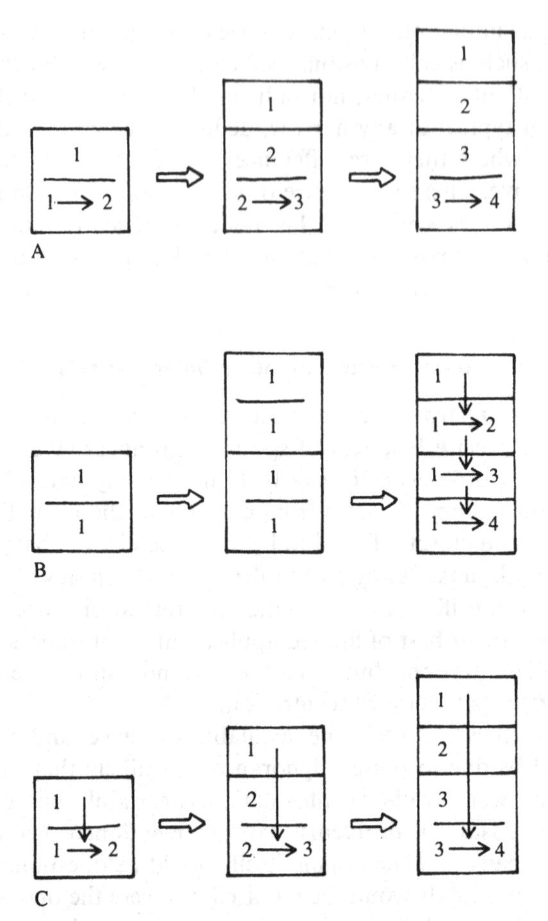

Figure 13.2. Integrated effects of intracellular and correlative controls on cell differentiation. A. The fate of cells can be due to internal counting mechanisms. Differentiation would occur during special, 'unequal' divisions (though this inequality need not be apparent). Such differentiation would be gradual. B. Cells differentiate in response to the correlative signals from other cells. A location at position '4' could be uniquely specified by signals coming from the three cellular types above. C. Combined, or 'integrated' effects of both intracellular and correlative controls of cell differentiation. Integration could occur if the response elicited by the correlative signals depended on previous differentiation, on a 'memory' of the signals the cells received during their past development.

continued increase of concentration or flow, or to continued supply to a region that uses the signal. These controls could also be much more reliable in specifying the fate of cells than any precise concentration of critical signals: small chance events would not have any lasting effects on development.

PATTERNING PRINCIPLES

The central topic of this book has not been the controls of differentiation but rather how these controls localize events so that they result in non-random, patterned tissues. Clearly, no one dominant patterning mechanism has been found. This is hardly surprising in view of the complexity of the tissues of plants or other living organisms. What is perhaps surprising is that the number of mechanisms found is not very large. The purpose of the following list is to serve both as a summary and as a basis for a discussion of broader principles of biological patterning.

(a) *Cell orientation and polarization*

The earliest stage in the pattern of plant tissues is the determination of a dominant cellular axis, one along which further growth and differentiation are oriented. This determination is a response to the passage of morphogenetic signals through the cells, a passage which occurs in all cells that are not in terminal locations. It follows that the initial difference between oriented and non-oriented cells could be a necessary consequence of growth and of the location of cells in terminal and non-terminal positions. It also follows that primary patterning information that cells obtain from their surrounding tissues concerns orientation (Chapters 5, 10).

The choice between the two possible directions along an axis, or the polarization of cells, may be a later stage of the same orienting processes (Chapter 6). At least for this later stage an important controlling signal, auxin, is chemically known (Chapters 3, 5). Furthermore, processes of polarization, or rather re-polarization, occur even in mature, non-meristematic cells (Chapter 5). The difference between the polarization reactions of embryonic and mature cells may be only quantitative: embryonic cells may become oriented much more readily, in response to shorter periods of signal passage.

(b) *Size determination by interactions with the rest of the plant*

The developmental fate of cells and groups of cells depends on limiting signals which they obtain from the rest of the plant (Chapter 2). This principle might determine the size of groups of similar cells. This could happen when groups of cells, such as promeristems (Chapter 10), have an advantage over one or a few cells. These promeristems must also have an advantage over larger regions, thus localizing development and preventing disorganization. The advantage could be due to the response of the plant – i.e., the limiting signals it provides – being greater than the sum of the signals provided to individual cells of the same state of differentiation. This would mean that cells can collaborate through their effects on the rest on the plant, and that direct, local interactions between neighboring cells are not always required. It could also mean that organized meristematic regions have an advantage which prevents the expression of the potential for unorganized growth, a potential present in most plant tissues (Chapter 4).

(c) *Gradients along the plant*

Gradients of signals or of cellular traits are due to two known causes. Because continued growth is at the root and shoot apices, the youngest or most embryonic cells are always at the tips – and the cells mature along a gradient that starts at these tips (Chapter 10). A second, unrelated gradient of cellular traits is due to the changes in the apical meristems, changes that are maintained in the mature cells they produce. These changes concern syndromes of traits along gradients from the juvenile to the non-juvenile or reproductive condition (Chapter 12). The differences along the plant axis may be gradual, extending along considerable distances, but they are not necessarily expressed in overt traits of tissue differentiation and organization, except during the formation of flowers. The changes in the apices appear to depend on the gradual accumulation of signals received from the rest of the plant.

(d) *Patterns due to intracellular determination*

As mentioned above, differentiation events can occur in a cell and be expressed in its products. This is straightforward when neighboring cells are similar because they are the products of a mother cell which underwent an early, stable differentiation. But actual patterning can occur within a mother cell, leading to unequal divisions, ones in which different but often complementary cell types are formed. The most prominent example of this is the one considered in Chapter 8, the development of stomata. It is possible that such processes occur where unequal divisions can not be readily observed, but the available evidence does not suggest that their occurrence is widespread.

(e) *Interactions between neighboring cells*

Cells differentiated in one way may induce similar or complementary differentiation in neighboring cells, leading to local patterns. The induction may well be reciprocal: there may be interactions and even feedback relations between the differentiation of neighboring cells. The signals passing from one cell to another could be formed locally and their effects could be limited to one or to very few neighboring cells. There is evidence for such inductive relations (in the development of stomata of monocotyledons, Chapter 8), but it is remarkable how meager this evidence is. There is better evidence for another type of interaction, in which the activity of the cells modifies the distribution of signals passing throughout the plant. The clearest example of this concerns the canalization of vascular differentiation to distinct rows of cells (Chapter 6). Cells through which signls pass become the preferred channels for the transport of the very same signals and this inhibits similar differentiation in neighboring cells from which the signals are drained by the efficient transport of the specialized cell. At the same time canalized signal flow induces similar differentiation in cells above and below the specialized cell. The signals involved in these interactions need be neither local nor specific to result in prevalent cellular patterns.

(f) *Packing of multicellular units*
Units that originate at random may be arranged in characteristic patterns when they are common enough to be in frequent contact with one another. When the units have complex forms and they occupy all available space (leaf primordia in phyllotaxis, Chapter 11), the patterns can be quite complex, especially so since they can be elaborated on by later growth. When the specialized units are composed of various cells (stomata, Chapter 8) the packing origin of the patterns need not be readily apparent.

Open problems and wild hypotheses

The general significance of this list of patterning mechanisms is discussed below. It is first desirable to ask whether, or to what extent, could these mechanisms be responsible for the main cellular patterns found in plants. In this context it is important that the patterning principles account for the extreme disruption of organization that characterizes tumors: the loss of the organised, limited formation of the long-distance signals, auxins and cytokinins, disrupts the relations between developmental centers (Chapter 4). And the same loss of organization also influences the controls of the size of promeristematic regions and causes the growth of unorganized callus even where meristematic centers are present (Chapter 10) as well as disrupting the polarity of the various tissues (Chapters 5, 6).

On the other hand, the limitation of the same patterning principles is seen in their inability, as yet, to offer clear insights concerning the organised development of insect galls (Fig. 13.3). In contrast to the bacteria that cause tumor formation, insects may influence the cells continuously, and their influence may even change with time, as the gall develops. Turning to the healthy, undamaged tissues, it is hardly surprising that not all cellular patterns have been considered – this would require a long catalogue of topics about which nothing concrete could be said. Thus the discussion above need not be wrong, but for obvious reasons it has dealt with highly simplified, perhaps unreal, plants. Major topics not covered include the presence of cones and sheets of cells of sclerenchyma or other tissues. Nor was there much that could be said about the quantitative relations between the various cell types along the radius of

Figure 13.3. Insect galls in *Pistacia palestina*: new organization determined by a foreign organism. The galls were caused by related aphids which settled on an entire leaf (A) or on a leaflet (B). The developmental changes were initiated when the leaves were very young but, judging from other insect galls, new development required the continued presence of the aphids.

the plant axis (Chapter 6) or the local polarization of the cells of leaf mesophyll.

Additional mechanisms required to account for these patterns include an additional polarity, along the radius of the stems and roots (Chapter 6). It is also likely that unknown signals are required (Chapter 3). But not all signals need be unknown substances. Physical stress, often a product of patterned growth, could orient further developmental events and be 'consumed' by appropriate growth responses (Chapter 11; Green, 1962, 1987; Hejnowicz, 1980; Green and Poethig, 1982). Electric fields are also an expected product of unequal development and there is good evidence that they could be a symptom of an early stage of cell polarization (Weisenseel et al., 1979; Jaffe, 1981). There are also various indications, though as yet no clear proof, that electrical phenomena may transmit developmental information (Pickard, 1973). Finally, oscillations or other orderly variations of the transport of substances could serve as developmental signals quite independently from the chemical nature of the substances themselves (Goodwin and Cohen, 1969; Hejnowicz, 1975). There is a detailed suggestion how known oscillations in the transport of one hormone, auxin, could result in a grid of 'positional information' (Hejnowicz, 1973; Zajaczkowski et al., 1983, 1984; Stieber, 1985). The short-coming of this suggestion is that it accounts only for domains of oriented events in the cambium, and even here the 'waves' would not specify the cell orientation that characterizes domains. This is not to deny the possibility that cellular patterns are related to interactions which change with time; this temporal organization is a most promising concept (Chapter 9).

TOWARDS A GENERAL THEORY OF BIOLOGICAL PATTERNING

Patterning in view of the available results

It is now necessary to re-state the general conceptual problem considered in this book – and to summarize the available answers, partial though they are. Discussions above dealt with examples of the problem of 'pattern formation' (Wolpert, 1971) as it applies to plant tissues. This is the problem of how chemical traits – the order of nucleotides in genes and the resulting structure of proteins – could specify microscopic form. To use an image, it is like the plan of a skyscraper – but where the plan is present in shape or hereditary information of the individual bricks, and originally all the bricks are uniform. The purpose now will be to seek generalizations valid beyond the tissues of seed plants or any other specific cases.

As mentioned in the Introduction, conceptual answers to this general problem of patterning are now available. These answers rely on early, chemical plans or pre-patterns and on intracellular programs of development, these being ways by which patterns could be specified separate from and preceding development itself. There is nothing in the previous chapters to contradict these concepts – but there is plenty to show that alone they could not suffice (Fig. 13.4). Thus, at a cellular level not only mature patterns but also the details of development have been found to be remarkably variable, perhaps more so than mature, functional structures. This variability is hardly expected if the fate of a cell is strictly specified at the time it is formed. Furthermore, there is concrete knowledge concerning controls of tissue organization, controls whose disruption leads to the development of tumors. Yet the signals involved – plant hormones – are neither specific to any single differentiation process nor do they act locally and at precise concentrations.

An examination of the patterning principles summarized in the previous section suggests ways in which available hypotheses of patterning might be extended. The roles of gradients of cellular traits, intracellular developmental programs and local interactions between cells are quite in accordance with accepted hypotheses. Less expected are the limited role of these principles, and the examples of patterning by processes in which the initial differentiation of cells can depend on chance. Where early events are partly random, patterns are imposed later, as development proceeds. Thus, the size of apical meristems could be determined after they are initiated, by a relation between size and competition for essential factors from the rest of the plant. Differentiation is canalized to distinct strands

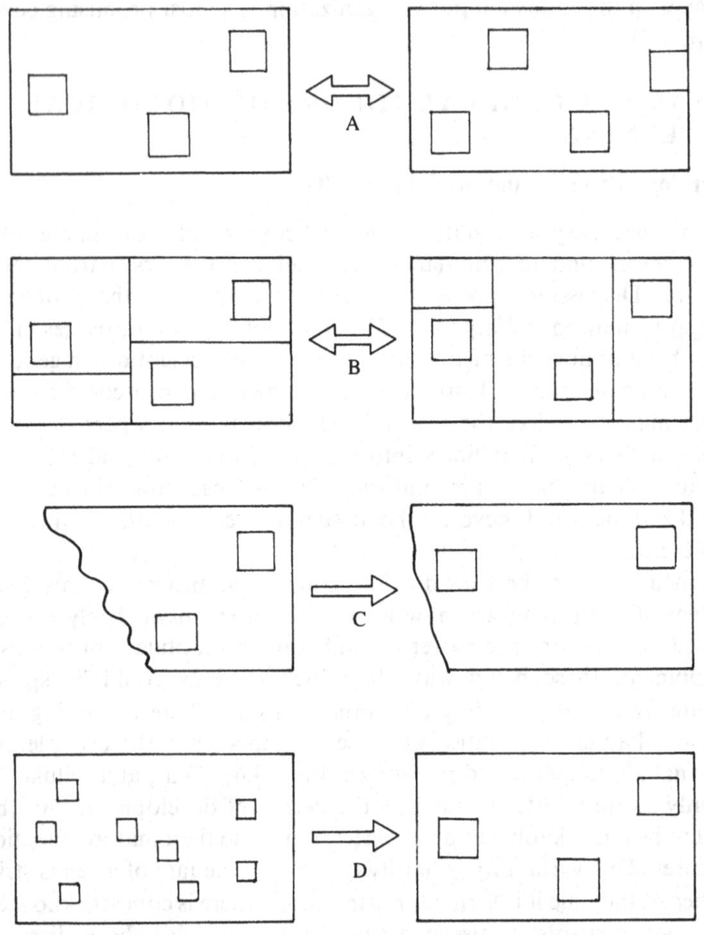

Figure 13.4. Indications that tissue patterning could not depend only on an early program of pre-pattern. A. Variability of cellular patterns. The two examples could come from identical locations in consecutive organs or in plants of the same genotype. Yet the number of specialized cells and their precise pattern are not the same. B. Variable development. The two examples could have come from comparable regions. Yet the parts formed from the four original cells, marked by lines, differ considerably for no apparent cause. C. Regeneration occurs even after the patterning of the tissue is apparent. Tissues damaged as shown on the left may reconstitute many features of their original structure. D. Differentiation processes may be initiated without leading to any functional structures. The early signs of differentiation may be found in many cells (on the left) and yet this differentiation is completed in only a few selected locations.

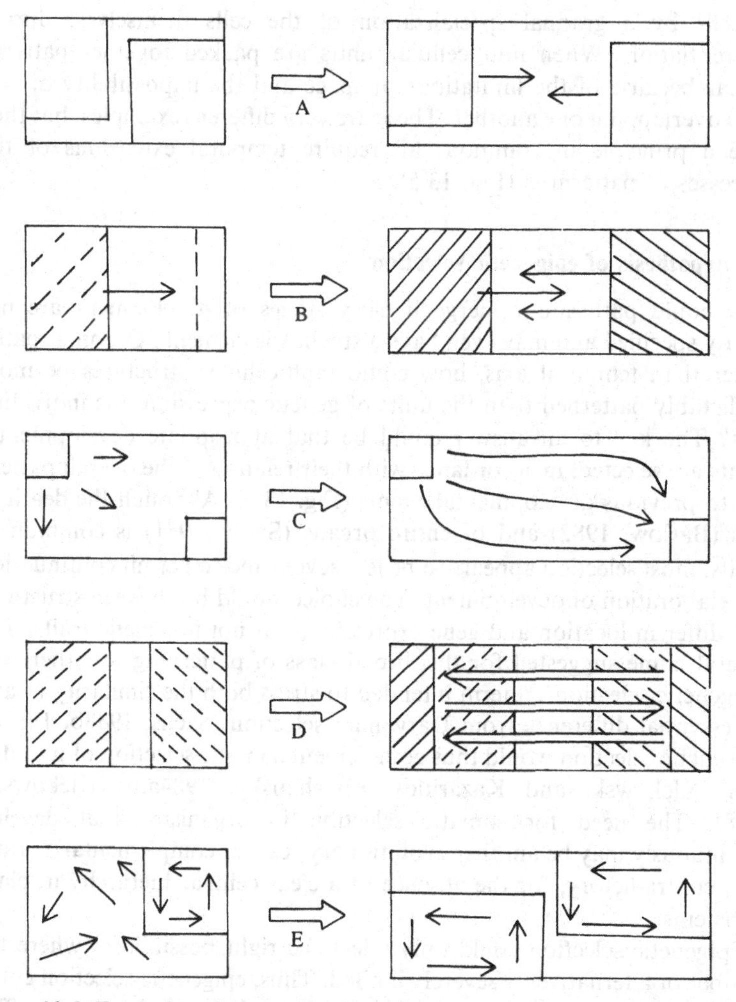

Figure 13.5. Examples of patterning during, rather than preceding, development A. Two multicellular units that are in close contact induce the growth of the intervening tissues. This growth separates these units, forming a new pattern. B. Mutual induction of differentiation. The specialized region on the left induces a complementary differentiation at the other end of the tissue. Inductive effects in the opposite direction lead gradually, during continued development, to a balance between the two complementary regions. C. Two regions induce the differentiation of specialized tissues between them. This induction specifies the orientation of the specialized tissues, not only their differentiation. D. Continued interactions of two differentiated regions led to their gradual specialization. This specialization means that the correlative interactions becomes limited to relatively small parts of the original regions. The ajoing tissues remain 'partially differentiated', a state which may be expressed by distinctive cell types. E. The 'packing' of multicellular units. The precise locations of the two units are not pre-determined, but the possibilities are severely constrained by available space.

of cells by a gradual specialization of the cells themselves, during differentiation. When multicellular units are packed together patterns appear because of the limitations of space and the impossibility of two units overlapping one another. These are very different examples, but they have a principle in common: all require temporal extensions of the processes of patterning (Fig. 13.5).

The hypothesis of epigenetic selection

How could patterning emerge if early stages of development are not strictly specified and may even have a stochastic element? Or, on a spatial rather than temporal axis, how could multicellular structures be more predictably patterned than the units of genetic expression, the individual cells? The key to an answer could be that appropriate developmental events are selected, in accordance with their relation to the overall pattern and to previous developmental events (Fig. 13.6). Although the death of cells (Barlow, 1982) and of entire organs (Snow, 1931) is common in plants, most selection appears to be less severe and to permit continuation and elaboration of development. The choice would be between structures that differ in location and gene expression, but not in genetic traits. The general name suggested for this broad class of patterning was therefore 'epigenetic selection', a term intended to stress both the similarity to and the essential difference from Darwinian selection (Sachs, 1988b, 1988c). Epigenetic selection would thus complement somatic selection of mutated cells (Klekowski and Kazarinova-Fukshansky, 1984a,b; Klekowski, 1988). The need for somatic selection in organisms that develop continuously may be another evolutionary reason, complementary rather than contradictory, for the absence of a clear cellular hierarchy in plant meristems.

Epigenetic selection could only select the right possibilities where the number of alternatives is severely limited. Thus, epigenetic selection could not act alone but rather in conjunction with developmental programs. The role of these programs, however, is often more limited than is commonly expected. Selection could depend on competition, so that the success of one process or structure excludes the occurrence or at least the continuation of others. In the examples above, success could mean the consumption or diversion of critical signals required for further development or the occupation of space. These examples suggest that organized, patterned development depends on limitations of the supplies of signals essential for differentiation. They further suggest that at least some of these signals act as coordinators of relatively large regions and are not merely concerned with relations between neighboring cells. The known facts concerning the disruption of plant tissues during the formation of tumors support these suggestions (Chapter 4).

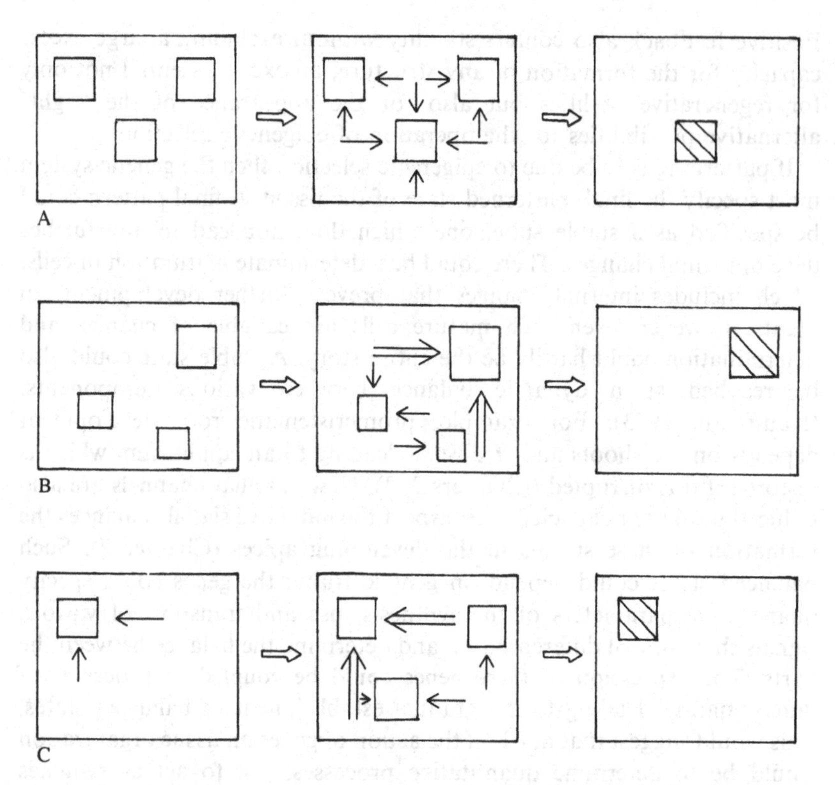

Figure 13.6. Selection – epigenetic selection – of appropriate structures from an excess of developmental possibilities. In all cases three initial structures compete. Only one of the three initials prevails and dominates the supply of differentiation-limiting signals, while the other initials mature as part of the cellular matrix. A. The relatively central location of one initial is an advantage that determines its further differentiation. B. One initial had a slightly enhanced rate of development, a difference that could be due to chance. The higher rate leads to a greater competitive ability and to the reversion of the differentiation of all other structures. C. The earliest initial prevails. It is larger than its competitors and this makes it a preferred sink for the signals that limit continued differentiation.

For competition and selection to lead to predictable results there must be criteria concerning the advantage of some processes over others. Relative position could be an important principle conferring such advantages. But in the examples of plant tissues considered above there are also advantages that depend on the time at which processes are initiated: processes that have already started tend to exclude those that would start later. There is thus a positive feedback between the occurrence of a process and its continuation, a feedback that tends to confer stability, or a conservative element, to development, and yet it does not exclude replacement or regeneration should internal or external conditions change.

Positive feedback also confers stability without excluding a large excess capacity for the formation of any structure, an excess essential not only for regenerative abilities but also for the appearance of the 'right' alternative possibilities for the operation of epigenetic selection.

If patterning is to be due to epigenetic selection then the genetic system must specify the final, patterned state of the tissue. A final pattern could be specified as a stable state, one which does not lead to any further developmental changes. There could be a determinate maturation of cells, which includes internal changes that prevent further development. In plants, however, even such mature cells are capable of change, and determination could hardly be the entire story. A stable state could also be reached as a dynamic balance between various components. (Kauffman, 1973). For example, promeristematic root development depends on the shoots and *vice versa*, leading to an equilibrium which is restored if it is disrupted (Chapters 2, 3). New vascular channels are also induced until their capacity to transport the inductive signals balances the formation of these signals in the developing apices (Chapter 7). Such balanced states could depend on genetic traits: the genes could specify quantitative parameters of the synthesis, use and transport of various signals that control differentiation and determine the balance between the parts. The expression of these genes could be coupled to processes of differentiation, leading to the gradual establishment of balanced states. This would suggest that much of the action of genes on tissue organization would be to determine quantitative processes, not to act as switches between alternative modes of gene expression. Indeed, homeotic genes that change one cell or tissue type to another are not known in plants – but it remains to be seen whether they are not known because they are rare, as suggested here, or because they have not been sought or are not viable. Perhaps it is also significant that although genes involved in development are now being isolated, there are as yet no clear indications of gene products that specify gradients of 'positional information'. Such genes have been identified in *Drosophila*, but they need not have the same role in plant development.

Epigenetic selection in relation to other hypotheses and other organisms

How different is epigenetic selection from other hypotheses of patterning? A first distinction is from hypotheses of strict programs in which each stage is a necessary consequence of previous conditions (Lindenmayer, 1984). Such programs, which probably have important roles, need not be complicated so as to specify elaborate patterns, since their effects can be superimposed, or integrated over time (Green and Poethig, 1982). Epigenetic selection also differs from hypotheses that the specification of patterns precedes their expression during development (Wolpert, 1971;

Mohr, 1978). Instead, in epigenetic selection patterning occurs during development and as a result of changes produced by processes of differentiation. These patterns are therefore 'differentiation-dependent' (Sachs, 1978a; Warren Wilson and Warren Wilson, 1984): they are due to the participation of the developing tissue, which forms, consumes and transports the critical signals of the developmental pre-pattern. The signals which are changed by development could include physical stresses, which both influence and are modified by growth (Chapter 11; Green, 1962, 1987; Barlow, 1984), and not only chemical substances that serve as developmental signals. Finally, epigenetic selection does not assume that the cells respond to precise concentrations of substances arranged in gradients (Lewis and Wolpert, 1976; Holder, 1979; Warren Wilson and Warren Wilson, 1984). Since the cells respond gradually, over an extended period, there is no requirement for a sensitivity to precise concentrations which could be an obvious source of developmental errors.

The possibilities of competition and selection can be readily found in earlier theories of biological patterning (Kirschner and Mitchison, 1986; Michaelson, 1987). Competition and selection are also the very heart of the formation of patterns by 'diffusion-reaction' (Meinhardt, 1982). In these models centers of a reaction continue, and reinforce themselves, only if they are not inhibited by similar, neighboring centers, an inhibition that can be due to competition for essential substrates. The formation of structures based on the selection of random events is also known from the inanimate world – it is the way crystals are formed in a homogeneous saturated solution. But there are differences between these suggestions and the one espoused here. Epigenetic selection may depend on chemical pre-patterns, but they are essentially dynamic pre-patterns, ones that changes and are 'corrected' during rather than preceding development. And this means that formation and consumption of signals need not be due to simple chemical processes and follow chemical kinetics; and transport need not be determined by diffusion and depend strongly on distance (Sachs, 1988b, c). Perhaps the processes that resemble epigenetic selection in the inanimate world are the long-term instabilities and the reversal of patterned processes in the formation of clouds and in the patterning of valleys by erosion.

Epigenetic selection could not be expected to lead to precise patterns, ones that are invariably repeated and optimally adapted to their mature functions. Nor would selection choose the shortest possible course of development. But as a developmental principle it has advantages that apparently compensate for this limitation of precision. Chief among the advantages of development by epigenetic selection may be the incorporation of compensation and even correction of developmental mistakes – and such mistakes are bound to occur in any complicated system. At the same time there is also the related incorporation of mechanisms of regulation or regeneration following damage, and this is not limited to early damage, before development actually occurs. Another advantage is the prevention rather than correction of mistakes: because the cells are required to respond to much more reliable signals, ones that have temporal dimensions. Finally, the specification of patterns by epigenetic selection is economical in terms of requirements for developmental rules and for special signals (Gordon, 1966).

Epigenetic selection was suggested on the basis of evidence concerning the patterning of plant tissues. Analogous though simpler processes could form inanimate patterns: clouds and valleys are the products of competing processes that interact with the pattern itself. What about animals? How general could the role of epigenetic selection be? Even in plant tissues no suggestion was made that selection acts alone but rather that it acts in conjunction with developmental programs. A role of epigenetic selection should therefore be sought in terms of relative contributions and these can be expected to vary from one group of organisms to another.

Indications of epigenetic selection could be variability at the cellular level and a large excess capacity for the formation of any given structure. There can also be many early initiations of any developmental process, not all of which are completed and some of which are even reversed. Development that exhibits these traits could not depend only on programs or pre-patterns, characterized by a strict specification of development before it actually occurs. And all these indications of epigenetic selection are common in animals. Cell lineages are extremely regular in nematodes (Sulston et al., 1983), but various degrees of variability, and even stochastic events (Iannaccone et al., 1987) are as common in animals as they are in seed plants. This variability of cell origin is stressed, rather than contradicted, by rules that show that variability can be limited, as in developmental compartments in *Drosophila* (Crick and Lawrence, 1975). A large excess capacity for forming a given structure is often revealed only by experimental manipulations. For example, in chick embryos wounds and grafts show graded capacities to form structures which normally form only in one place (Khaner and Eyal-Giladi, 1986). Finally, a reversal of processes once they have occurred is common in the formation of nerve connections (Changeux and Danchin, 1976) – and here the 'best' contacts are actually selected, in accordance with conditions during development rather than those preceding it.

Epigenetic selection appears to be a common process. Could one generalize about its relative contribution in different organisms? The apical meristems of seed plants have a large stochastic element at the cellular level (Chapter 7), while simpler plants have apical cells that follow a strict program of oriented divisions. This suggests that the developmental role of epigenetic selection increases with the complexity of the organism. The evidence from animals would tend to agree with this suggestion – nematodes, the best example of programmed development, are certainly simpler than vertebrates where lineages have a stochastic element. It would make sense for complex organisms to require the compensation or correction supplied by epigenetic selection. On the other hand, relatively simple organisms often have remarkable regenerative abilities, indicating that they have active morphogenetic fields even in the mature state. Perhaps selection mechanisms are a general rule and it is the stochastic element that is more pronounced in complicated organisms. A wide, comparative study of patterning in different organisms is clearly required.

References

Addicott, F. T. (1970). Plant hormones and their role in abscission. *Biological Reviews of the Cambridge Philosophical Society* 45: 485–524.

Albersheim, P. & Darvill, A. G. (1985). Oligosaccharins. *Scientific American* 253 (3), September 44–50.

Allsopp, A. (1964). Shoot morphogenesis. *Annual Review of Plant Physiology* 15: 225–54.

Aloni, R. (1976). Polarity of induction and pattern of primary phloem fiber differentiation in *Coleus. American Journal of Botany* 63: 877–89.

Aloni, R. (1978). Source of induction and sites of primary phloem fibre differentiation in *Coleus blumei. Annals of Botany* 42: 1261–9.

Aloni, R. (1979). Role of auxin and gibberellin in the differentiation of primary phloem fibers. *Plant Physiology* 63: 609–14.

Aloni, R. (1980). Role of auxin and sucrose in the differentiation of sieve and tracheary elements in plant tissue cultures. *Planta* 150: 255–63.

Aloni, R. (1987a). The induction of vascular tissues by auxin. In *Plant Hormones and their Role in Plant Growth and Development*, ed. P. J. Davies, pp. 363–74. Dordrecht: Martinus Nijhoff.

Aloni, R. (1987b). Differentiation of vascular tissues. *Annual Review of Plant Physiology* 38: 179–204.

Aloni, R. (1988). Vascular differentiation within the plant. Chapter 3 in *Vascular Differentiation and Plant Growth Regulators*, by L. W. Roberts, P. B. Gahan & R. Aloni, pp. 39–62. Berlin: Springer.

Aloni, R. & Sachs, T. (1973). The three dimensional structure of primary phloem systems. *Planta* 113: 345–53.

Amasino, R. M. & Miller, C. O. (1982). Hormonal control of tobacco crown gall morphology. *Plant Physiology* 69: 389–92.

Arney, S. E. (1956). Studies on the growth and development of the genus *Fragaria*. VIII. The effect of defoliation on leaf initiation and early growth of leaf initials. *Φyton* 6: 109–20.

Avery, G. S., Jr. (1935). Differential distribution of a phytohormone in a developing leaf of *Nicotiana*, and its relation to polarized growth. *Bulletin of the Torrey Botanical Club* 62: 313–30.

Balatinecz, J. J. & Farrar, J. L. (1966). Pattern of renewed cambial activity in relation to exogenous auxin in detached woody shoots. *Canadian Journal of Botany* 44: 1108–10.

Ball, E. (1952). Experimental division of the shoot apex of *Lupinus albus* L. *Growth* 16: 151–74.

Ball, E. (1960). Cell divisions in living shoot apices. *Phytomorphology* 10: 377–96.

Banks, M. S. (1979). Plant regeneration from callus from two growth phases of English ivy, *Hedera helix* L. *Zeitschrift für Pflanzenphysiologie* 92: 349–53.

Bannan, M. W. (1951). The annual cycle of changes in the fusiform cambial cells of *Chamaecyparis* and *Thuja*. *Canadian Journal of Botany* 29: 421–37.
Baranova, M. A. (1987). Historical development of the present classification of morphological types of stomates. *The Botanical Review* 53: 53–79.
Barlow, P. W. (1975). The root cap. In *The Development and Function of Roots*, eds. J. G. Torrey & D. T. Clarkson, pp. 21–54. London: Academic Press.
Barlow, P. W. (1976). Towards an understanding of the behaviour of root meristems. *Journal of Theoretical Biology* 57: 433–51.
Barlow, P. W. (1978). The concept of stem cell in the context of plant growth and development. In *Stem Cells and Tissue Homeostasis*, eds. B. I. Lord, C. S. Potten & R. J. Cole, pp. 87–113. Cambridge: Cambridge University Press.
Barlow, P. W. (1981). Division and differentiation at the root apex. In *Structure and Function of Plant Roots*, eds. R. Brouwer, O. Gašparíková, J. Kolek & B. C. Loughman, pp. 85–7. The Hague: M. Nijhoff/W. Junk.
Barlow, P. W. (1982). Cell death – an integral part of plant development. In *Growth Regulators in Plant Senscence*, eds. M. B. Jackson, B. Grout & I. A. Mackenzie, (Monograph 8 of the British Plant Growth Regulator Group), pp. 27–45. Wantage, British Plant Growth Regulator Group.
Barlow, P. W. (1984). Positional controls in root development. In *Positional Controls in Plant Development*, eds. P. W. Barlow & D. J. Carr, pp. 281–318. Cambridge: Cambridge University Press.
Barlow, P. W. (1986). Adventitious roots of whole plants: their forms, functions and evolution. In *New Root Formation in Plants and Cuttings*, ed. M. B. Jackson, pp. 67–110. Dordrecht: Martinus Nijhoff.
Barlow, P. W. (1987). Requirements for hormone involvement in development at different levels of organization. In *Hormone Action in Plant Development: A Critical Appraisal*, eds. S. G. V. Hoad, J. R. Lenton, M. B. Jackson & R. K. Atkin, pp. 39–51. London: Butterworth.
Barlow, P. W. & Adam, J. S. (1988). The position and growth of lateral roots on cultured root axes of tomato, *Lycopersicon esculentum* (Solanaceae). *Plant Systematics and Evolution* 158: 141–54.
Barlow, P. W. & Carr, D. J. (eds.) (1984). *Positional Controls in Plant Development*. Cambridge: Cambridge University Press.
Barlow, P. W. & Rathfelder, E. L. (1985). Cell division and regeneration in primary root meristems of *Zea mays* recovering from cold treatment. *Environmental and Experimental Botany* 25: 303–14.
Barlow, P. W. & Sargent, J. A. (1978). The ultrastructure of the regenerating root cap of *Zea mays* L. *Annals of Botany* 42: 791–9.
Barnett, J. R. (ed.) (1981). *Xylem Cell Development*. Tunbridge Wells, Kent: Castle House.
Battey, N. H. & Lyndon, R. F. (1986). Apical growth and modification of the development of primordia during re-flowering of reverted plants of *Impatiens balsamina* L. *Annals of Botany* 58: 333–41.
Beal, W. J. (1873). Phyllotaxis of cones. *American Naturalist* 7: 449–53.
Beardsell, D. V. & Considine, J. A. (1987). Lineage, lineage stability and pattern formation in leaves of variegated chimeras of *Lophostemon confertus* (R. Br.) Wilson and Waterhouse and *Tristaniopsis lavrina* (Smith) Wilson and Waterhouse (Myrtaceae). *Australian Journal of Botany* 35: 701–14.
Benayoun, J., Aloni, R. & Sachs, T. (1975). Regeneration around wounds and the control of vascular differentiation. *Annals of Botany* 39: 447–54.
Bergfeld, R., Speth, V. & Schopfer, P. (1988). Reorietation of microfibrils and microtubules at the outer epidermal wall of maize coleoptiles during auxin mediated growth. *Botanica Acta* 101: 57–67.

Bernier, G., Kinet, J. M. & Sachs, R. M. (1981). *The Physiology of Flowering*, Volumes 1 and 2. Boca Raton, Florida: C.R.C. Press.

Bienick, M. E. & Millington, W. F. (1967). Differentiation of lateral shoots as thorns in *Ulex europaeus. American Journal of Botany* 54: 67–70.

Bierhorst, D. W. (1977). On the stem apex, leaf initiation and early leaf ontogeny of filicalean ferns. *American Journal of Botany* 64: 125–52.

Binding, H., Witt, D., Monzer, J., Mordhorst, G. & Kollmann, R. (1987). Plant cell graft chimeras obtained by co-culture of isolated protoplasts. *Protoplasma* 141: 64–73.

Blackman, P. G. & Davies, W. J. (1985). Root to shoot communication in maize plants of the effects of soil drying. *Journal of Experimental Botany* 36: 39–48.

Bloch, R. (1965). Polarity and gradients in plants: a survey. *Encyclopedia of Plant Physiology*, ed. W. Ruhland, vol. 15 (1), pp. 234–74. Berlin: Springer.

Bode, P. M. & Bode, H. R. (1984). Patterning in *Hydra*. In *Pattern Formation*, eds. G. M. Malacinski, & S. V. Bryant, pp. 213–41. New York: Macmillan.

Bonner, J. (1942). Transport of thiamine in the tomato plant. *American Journal of Botany* 29: 136–42.

Bonner, J. & English, J., Jr. (1938). A chemical and physiological study of traumatin, a plant wound hormone. *Plant Physiology* 13: 331–48.

Bourbouloux, A. & Bonnemain, J. L. (1979). The different components of the movement and the areas of retention of labeled molecules after the application of (^{3}H)indolyl-acetic acid to the apical bud of *Vicia faba. Physiologia Plantarum* 47: 260–8.

Braun, A. C. (1953). Bacterial and host factors concerned in determining tumor morphology in crown gall. *Botanical Gazette* 104: 363–71.

Braun, A. C. (1956). The activation of two growth substance systems accompanying the conversion of normal to tumor cells in crown gall. *Cancer Research* 16: 53–6.

Braun, A. C. (1958). The physiological basis for the autonomous growth of the crown gall tumor. *Proceedings of the National Academy of Sciences of the U.S.A.* 44: 344–9.

Braun, A. C. (1978). Plant tumors. *Biochimica et Biophysica Acta* 516: 167–91.

Brawley, S. H., Wetherell, D. F. & Robinson, K. R. (1984). Electrical polarity in embryos of wild carrot precedes cotyledon differentiation. *Proceedings of the National Academy of Sciences of the U.S.A.* 81: 6064–7.

Brenner, M. L. (1987). The role of hormones in photosynthetic partitioning and seed filling. In *Plant Hormones and their Role in Plant Growth and Development*, ed. P. J. Davies, pp. 474–93. Dordrecht: Martinus Nijhoff.

Brenner, S., Murray, J. D. & Wolpert, L. (eds.) (1981). Theories of biological pattern formation. *Philosophical Transactions of the Royal Society of London*, B 295: 425–617.

Bruck, D. K. & Paolillo, D. J., Jr. (1984a). The control of vascular branching in *Coleus*. I. The side bundle. *Annals of Botany* 53: 727–36.

Bruck, D. K. & Paolillo, D. J., Jr. (1984b). The control of vascular branching in *Coleus*. II. The corner traces. *Annals of Botany* 53: 737–47.

Bruck, D. K. & Walker, D. B. (1985). Cell determination during embryogenesis in *Citrus jambhiri*. I. Ontogeny of the epidermis. *Botanical Gazette* 146: 188–95.

Bünning, E. (1951). Über die Differezierungsvorgängen in der Cruciferenwurzel. *Planta* 39: 126–53.

Bünning, E. (1953). *Entwicklungs- und Bewegungsphysiologie der Pflanze*, 3rd edn. Berlin: Springer-Verlag.

Bünning, E. (1957). Polarität und inäquale Teilung des Pflanzlichen Protoplasten. *Protoplasmatologia* 8(9a): 1–86.

Bünning, E. (1965). Die Entstehung von Mustern in der Entwicklung von Pflanzen. *Encyclopedia of Plant Physiology*, ed. W. Ruhland, vol. 15(1), pp. 383–408. Berlin: Springer.
Bünning, E. & Ilg, H. (1954). Polaritätsstörungen bei Pflanzenzellen durch Äthylen. *Planta*, 43: 472–6.
Bünning, E. & Sagromsky, H. (1948). Die Bildung des Spaltöffnungsmusters in der Blattepidermis. *Zeitschrift für Naturforschung* 3b: 203–16.
Burgess, J. (1972). The occurrence of plasmodesma-like structures in a non-division wall. *Protoplasma* 74: 449–58.
Burgess, J. (1985). *An Introduction to Plant Cell Development*. Cambridge: Cambridge University Press.
Burgess, J. & Linstead, P. (1984). In vitro tracheary element formation: structural studies and the effect of tri-iodobenzoic acid. *Planta* 160: 481–9.
Buvat, R. (1955). Le méristème apical de la tige. *Anneé Biologique, 3ème série*, 31: 595–656.
Camus, G. (1949). Recherches sur le rôle des bourgeons dans les phénomènes de morphogénèse. *Revue de Cytologie et de Biologie Végétale* 11: 1–195.
Carmi, A. & Heuer, B. (1981). The role of roots in control of bean shoot growth. *Annals of Botany* 48: 519–27.
Carmi, A., Sachs, T. & Fahn, A. (1972). The relation of ray spacing to cambial growth. *The New Phytologist* 71: 349–53.
Carr, D. J. (1984). Positional information in the specificjation of leaf, flower and branch arrangement. In *Positional Controls in Plant Development*, eds. P. W. Barlow & D. J. Carr, pp. 441–60. Cambridge: Cambridge University Press.
Chandorkar, K. R. & Dengler, N. G. (1987). Effect of low level continuous gamma irradiation on vascular cambium activity in Scotch pine, *Pinus sylvestris* L. *Environmental and Experimental Botany* 27: 165–75.
Changeux, J.-P. & Danchin, A. (1976). Selective stabilization of developing synapses as a mechanism for the specification of neuronal networks. *Nature* 264: 705–12.
Charlton, W. A. (1974). Studies in the *Alismataceae*. V. Experimental modification of phyllotaxis in pseudostolons of *Echinodorus tenellus* by means of growth inhibitors. *Canadian Journal of Botany* 52: 1131–42.
Charlton, W. A. (1982). Distribution of lateral root primordia in the root tips of *Musa acuminata* Colla. *Annals of Botany* 49: 509–20.
Charlton, W. A. (1987). Relationship between lateral root primordia in different ranks. *Annals of Botany* 60: 455–8.
Charlton, W. A. (1988). Stomatal pattern in four species of monocotyledons. *Annals of Botany* 61: 611–21.
Child, C. M. (1941). *Patterns and Problems of Development*. Chicago: University of Chicago Press.
Chibnall, A. C. (1954). Protein metabolism in rooted runner bean leaves. *The New Phytologist* 53: 31–7.
Chouard, P. & Aghion, D. (1961). Modalités de la formation de bourgeons floraux sur des cultures de segments de tige de Tabac. *Comptes Rendus hebdomadaires des Séances de l'Académie des Sciences, Paris*, 252: 3864–6.
Christianson, M. L. (1986). Fate map of the organizing shoot apex in *Gossypium*. *American Journal of Botany* 73: 947–58.
Christou, P. (1988). Habituation in *in vitro* soybean cultures. *Plant Physiology* 87: 809–12.
Church, A. H. (1904). *On the Relation of Phyllotaxis to Mechanical Laws*. London: Williams & Norgate.

Clowes, F. A. L. (1959). Apical meristems of roots. *Biological Reviews of the Cambridge Philosophical Society* 34: 501–29.
Clowes, F. A. L. (1961). *Apical Meristems*. Oxford: Blackwell.
Clowes, F. A. L. (1984). Size and activity of quiescent centres in roots. *The New Phytologist* 96: 13–21.
Clowes, F. A. L. & MacDonald, M. M. (1987). Cell cycling and the fate of potato buds. *Annals of Botany* 59: 141–8.
Coe, E. H., Jr. & Neuffer, M. G. (1978). Embryo cells and their destinies in the corn plant. In *The Clonal Basis of Development*, eds. S. Subtelny & I. M. Sussex, pp. 113–29. New York: Academic Press.
Corner, E. J. H. (1958). Transference of function. *Journal of the Linnean Society of London* 56: 33–40.
Cornish, E. C., Anderson, M. A. & Clarke, A. E. (1988). Molecular aspects of fertilization in flowering plants. *Annual Review of Cell Biology* 4: 209–28.
Crick, F. H. C. (1970). Diffusion in embryogenesis. *Nature* 225: 420–2.
Crick, F. H. C. & Lawrence, P. A. (1975). Compartments and polyclones in insect development. *Science* 189: 340–7.
Cunninghame, M. E. & Lyndon, R. F. (1986). The relationship between the distribution of periclinal cell divisions in the shoot apex and leaf initiation. *Annals of Botany* 57: 737–46.
Cusick, F. (1956). Studies on floral morphogenesis. I. Median bisections of flower primordia in *Primula bullyana* Forrest. *Transactions of the Royal Society, Edinburgh* 63: 153–66.
Cusset, G. (1986). La morphogenèse du limbe des Dicotylédones. *Canadian Journal of Botany* 64: 2807–39.
Cutter, E. G. (1955). Experimental and analytical studies of pteridophytes XXIX. The effect of progressive starvation on the growth and organization of the shoot apex of *Dryopteris aristata*. *Annals of Botany* 29: 485–99.
Cutter, E. G. (1956). Experimental and analytical studies of pteridophytes. XXXIII. The experimental induction of buds from leaf primordia in *Dryopteris aristata* Druce. *Annals of Botany* 20: 143–65.
Cutter, E. G. (1959). On a theory of phyllotaxis and histogenesis. *Biological Reviews of the Cambridge Philosophical Society* 34: 243–63.
Cutter, E. G. (1964). Phyllotaxis and apical growth. *The New Phytologist* 63: 39–46.
Cutter, E. G. (1965). Recent experimental studies of the shoot apex and shoot morphogenesis. *The Botanical Review* 31: 7–113.
Cutter, E. G. & Feldman, L. J. (1956). Trichoblasts in *Hydrocharis*. I. Origin, differentiation, dimensions and growth. *American Journal of Botany* 57: 190–201.
Czaja, A. Th. (1935). Polarität und Wuchsstoff. *Berichte der deutschen botanischen Gesellschaft* 53: 197–220.
Davies, P. J. (1987a). The plant hormones: their nature, occurrence, and functions. In *Plant Hormones and their Role in Plant Growth and Development*, ed. P. J. Davies, pp. 1–11. Dordrecht: Martinus Nijhoff.
Davies, P. J. (1987b). The plant hormones concept: transport, concentration and sensitivity. In *Plant Hormones and their Role in Plant Growth and Development*, ed. P. J. Davies, pp. 12–23. Dordrecht: Martinus Nijhoff.
De Bary, A. (1877). *Vergleichende Anatomie der Vegetationsorgane der Phanerogamen und Farne*. Leipzig: Verlag Wilhelm Engelmann.
DeRopp, R. S. (1947). The growth promoting and tumefascient factors of bacteria-free crown-gall tumor tissue. *American Journal of Botany* 34: 248–61.
DeRopp, R. S. (1951). The crown gall problem. *The Botanical Review* 17: 629–70.

Desbiez, M. O., Kergosian, Y., Champagnat, P. & Thellier, M. (1984). Memorization and delayed expression of regulatory messages in plants. *Planta* 160: 392–9.
Dostál, R. (1909). Die Korrelationsbeziehung zwischen dem Blatt und seiner Axillarknospe. *Berichte der deutschen botanischen Gesellschaft* 27: 547–54.
Dostál, R. (1967). *On Integration in Plants*. Cambridge, Mass.: Harvard University Press.
Dravnieks, D. E., Skoog, F. & Burris, R. H. (1969). Cytokinin activation of de novo thiamine synthesis in tobacco callus cultures. *Plant Physiology* 44: 866–70.
Drew, A. P. (1982). Shoot–root plasticity and episodic growth in red pine seedlings. *Annals of Botany* 49: 347–57.
Dulieu, H. (1968). Emploi des chimères chlorophyliennes pour l'étude de l'ontogenie foliare. *Bulletin des Sciences, Bourgogne* 25: 13–72.
Einset, J. W. (1977). Two effects of cytokinins on the auxin requirement of tobacco callus cultures. *Plant Physiology* 59: 45–7.
Erickson, R. O. (1983). The geometry of phyllotaxis. In *The Growth of Leaves*, eds. J. E. Dale & F. L. Milthorpe, pp. 53–88. Cambridge: Cambridge University Press.
Erwee, M. G. & Goodwin, P. B. (1983a). Characterization of the *Egeria densa* Planch. leaf symplast. Inhibition of the intercellular movement of fluorescent probes by group II ions. *Planta* 158: 320–8.
Erwee, M. G. & Goodwin, P. B. (1983b). Characterization of the *Egeria densa* Planch. leaf symplast: response to plasmolysis, deplasmolysis and the aromatic amino acids. *Protoplasma* 122: 162–8.
Erwee, M. G. & Goodwin, P. B. (1985). Symplast domains in extrastelar tissues of *Egeria densa* Planch. *Planta* 163: 9–19.
Erwee, M. G., Goodwin, P. B. & Van Bell, A. J. E. (1985). Cell–cell communication in the leaves of *Commelina cyanea* and other plants. *Plant, Cell and Environment* 8: 173–8.
Esau, K. (1965). *Vascular Differentiation in Plants*. New York: Holt, Rinehart & Winston.
Esau, K. (1977). *Anatomy of Seed Plants*, 2nd edn. New York: Holt, Rinehart & Winston.
Eschrich, W. (1954). Beiträge zur Kenntnis die Wundsiebröhrentwicklung bei *Impatiens holstii*. *Planta* 43: 37–74.
Evans, L. T. (1960). Inflorescence initiation in *Lolium temulentum* L. I. Effect of plant age and leaf area on sensitivity to photoperiodic induction. *Australian Journal of Biological Sciences* 13: 123–31.
Evert, R. F. (1961). Some aspects of cambial development in *Pyrus communis*. *American Journal of Botany* 48: 479–88.
Fahn, A. (1982). *Plant Anatomy*, 3rd edn. Oxford: Pergamon Press.
Feldman, L. J. (1977). The generation and elaboration of primary vascular tissues patterns in roots of *Zea*. *Botanical Gazette* 138: 393–401.
Feldman, L. J. (1979). Cytokinin biosynthesis in the roots of corn. *Planta* 145: 315–21.
Feldman, L. J. (1984). The development and dynamics of the root apical meristem. *American Journal of Botany* 71: 1308–14.
Fisher, D. G. (1988). Movement of Lucifer Yellow in leaves of *Coleus Blumei* Benth. *Plant, Cell and Environment* 11: 639–44.
Fosket, D. E. (1972). Meristematic activity in relation to wound xylem differentiation. *Advances in Experimental Medicine and Biology* 18: 33–50.
Foster, A. S. (1941). Comparative studies on the structure of the shoot apex in seed plants. *Bulletin of the Torrey Botanical Club* 68: 339–50.

Francis, D. & Lyndon, R. F. (1979). Synchronization of cell division in the shoot apex of *Silene* in relation to flower initiation. *Planta* 145: 151–7.
French, V., Bryant, P. J. & Bryant, S. V. (1976). Pattern regulation in epimorphic fields. *Science* 193: 969–81.
Fry, S. C. & Wangermann, E. (1976). Polar transport of auxin through embryos. *The New Phytologist* 77: 313–17.
Frydman, V. M. & Wareing, P. F. (1973). Phase change in *Hedera helix* L. II The possible role of the roots as a possible source of shoot gibberellin-like substances. *Journal of Experimental Botany* 24: 1139–48.
Gahan, P. B. (1981). An early cytochemical marker of commitment to stelar differentiation in meristems from dicotyledonous plants. *Annals of Botany* 48: 769–75.
Gahan, P. B. (1988). Xylem and phloem differentiation in perspective. Chapter 1 in *Vascular Differentiation and Plant Growth Regulators*, by L. W. Roberts, P. B. Gahan & R. Aloni, pp. 1–21. Berlin: Springer.
Gautheret, R. J. (1944). Recherches sur la polarité des tissus végétaux *Revue de Cytologie et de Cytophysiologie Végétale* 7: 45–217.
Gautheret, R. J. (1947). Sur les besoins en hétéro-auxine des cultures de tissus de quelques végétaux. *Comptes Rendus de la Société de Biologie et de ses Filiales* 141: 627–9.
Gautheret, R. J. (1957). Histogenesis in plant tissue cultures. *Journal of the National Cancer Institute* 19: 555–73.
Gautheret, R. J. (1959). *La Culture des Tissus Végétaux*. Paris: Masson et Cie.
Gautheret, R. J. (1983). Plant tissue cultures: a history. *Botanical Magazine (Tokyo)* 96: 393–410.
Gersani, M. (1985). Appearance of transport capacity in wounded plants. *Journal of Experimental Botany* 36: 1809–16.
Gersani, M., Leshem, B. & Sachs, T. (1986). Impaired polarity in abnormal plant development. *Journal Plant Physiology* 123: 91–5.
Gersani, M., Lips, S. H. & Sachs, T. (1980a). The influence of roots, shoots and hormones on sucrose distribution. *Journal of Experimental Botany* 31: 177–84.
Gersani, M., Lips, S. H. & Sachs, T. (1980b). The influence of roots, shoots and hormones on the distribution of leucine, phosphate and benzyladenine. *Journal of Experimental Botany* 31: 777–82.
Gersani, M. & Sachs, T. (1984). Polarity reorientation in beans expressed by vascular differentiation and polar auxin transport. *Differentiation* 25: 205–8.
Gierer, A. & Meinhardt, H. (1972). Theory of biological pattern formation. *Kybernetik* 12: 30–9.
Gifford, E. M. (1954). The shoot apex in angiosperms. *The Botanical Review* 20: 477–529.
Gifford, E. M. (1983). Concept of apical cells in bryophytes and pteridophytes. *Annual Review of Plant Physiology* 34: 419–40.
Gifford, E. M. & Corson, G. E. (1971). The shoot apex in seed plants. *The Botanical Review* 37: 143–229.
Gifford, E. M. & Tepper, H. B. (1962). Ontogenetic and histochemical changes in the vegetative shoot tip of *Chenopodium album*. *American Journal of Botany* 49: 902–11.
Goebel, K. (1900). *Organography in Plants. I. General Organography*, translated by I. B. Balfour. Oxford: At the Clarendon Press.
Goebel, K. (1922). *Gesetzmässigkeiten in Blattaufbau.* (Botanisches Abhandlungen, Heft 1, 1–78). Jena: Fischer.

Goldberg, R. B. (1988). Plants: novel developmental processes. *Science* 240: 1460–7.
Goldsmith, M. H. M. (1977). The polar transport of auxin. *Annual Review of Plant Physiology* 28: 439–78.
Goodwin, B. C. & Cohen, M. H. (1969). A phase shift model for the spatial and temporal organization of developing systems. *Journal of Theoretical Biology* 25: 49–107.
Goodwin, P. B. (1983). Molecular size limit for movement in the symplast of *Elodea* leaf. *Planta* 157: 124–30.
Goodwin, P. B., Gollnow, B. I. & Letham, D. S. (1978). Phytohormones and growth correlations. In *Phytohormones and Related Compounds, A Comprehensive Treatise*, eds. D. S. Letham, P. B. Goodwin & T. J. V. Higgins, vol. 2, pp. 215–49. Amsterdam: Elsevier/North Holland.
Goosen-de-Roo, L. (1973). The structure of protoplast in primary tracheary elements of the cucumber after plasmolysis. *Acta Botanica Neerlandica* 22: 467–85.
Gopal, B. V. & Shah, G. L. (1970). Observations on normal and abnormal stomatal features in four species of *Asparagus* L. *American Journal of Botany* 57: 665–9.
Gordon, R. (1966). On stochastic growth and form. *Proceedings of the National Academy of Sciences of the U.S.A.* 56: 1497–504.
Gorst, J., Overall, R. L. & Wernicke, W. (1987). Ionic currents traversing cell clusters from carrot suspension cultures reveal perpetuation of morphogenetic potential as distinct from induction of embryogenesis. *Cell Differentiation* 21: 101–9.
Gorter, C. J. (1965). Origin of fasciations. *Encyclopedia of Plant Physiology*, ed. W. Ruhland, vol. 15(2), pp. 330–51. Berlin: Springer.
Gottlieb, L. D. (1986). The genetic basis of plant form. *Philosophical Transactions of the Royal Society, London*, B 313: 197–208.
Gould, S. J. (1977). *Ontogeny and Phylogeny*. Cambridge, Mass.: Harvard University Press.
Green, P. B. (1962). Mechanism for plant cellular morphogenesis. *Science* 138: 1404–5.
Green, P. B. (1976). Growth and cell pattern formation on an axis: critique of concepts, terminology and modes of study. *Botanical Gazette* 137: 187–202.
Green, P. B. (1980). Organogenesis: a biophysical view. *Annual Review of Plant Physiology* 31: 51–82.
Green, P. B. (1984). Shifts in plant cell axiality: histogenetic influences on cellulose orientation in the succulent *Graptopetalum*. *Developmental Biology* 103: 18–27.
Green, P. B. (1985). The surface of the shoot apex: a reinforcement-field theory for phyllotaxis. *Journal of Cell Science*, Supplement 2, 181–201.
Green, P. B. (1986). Plasticity in shoot development: a biophysical view. *Symposia of the Society for Experimental Biology* 40: 211–32.
Green, P. B. (1987). Inheritance of pattern: analysis from phenotype to gene. *American Zoologist* 27: 657–73.
Green, P. B. (1988). A theory of inflorescence development and flower formation based on morphological and biophysical analysis in *Echeveria*. *Planta* 175: 153–69.
Green, P. B. & Baxter, D. R. (1987). Phyllotactic patterns: characterization by geometrical activity at the formative region. *Journal of Theoretical Biology* 128: 387–95.
Green, P. B. & Poethig, R. S. (1982). Biophysics of extension and initiation of

plant organs. In *Developmental Order: Its Origin and Regulation*, eds. P. B. Green & S. Subtelny, pp. 485–509. New York: Alan R. Liss.

Gregory, F. C. & Veale, J. A. (1957). A reassessment of the problem of apical dominance. *Symposia of the Society for Experimental Biology* 11: 1–20.

Gresshoff, P. M., Skotnicki, M. L. & Rolfe, B. G. (1979). Crown gall teratoma formation is plasmid and plant controlled. *Journal of Bacteriology* 137: 1020–1.

Guern, J. (1987). Regulation from within: the hormone dilemma. *Annals of Botany* 60, Supplement 4: 75–102.

Gulline, H. F. & Walker, R. (1957). The regeneration of severed pea apices. *Australian Journal of Botany* 5: 129–36.

Gunning, B. E. S. (1978). Age related and origin control of the numbers of plasmodesmata in the cell wall of developing *Azolla* roots. *Planta* 143: 181–90.

Gunning, B. E. S. (1982). The root of the water fern *Azolla*: cellular basis of development and multiple roles for cortical microtubules. In *Developmental Order: Its Origin and Regulation*, eds S. Subtelny & P. B. Green, pp. 379–421. New York: A. R. Liss.

Gunning, B. E. S. & Barkley, W. K. (1963). Kinin-induced directed transport and senescence in detached leaves. *Nature* 199: 262–5.

Gunning, B. E. S. & Hardham, A. R. (1982). Microtubules. *Annual Review of Plant Physiology* 33: 652–98.

Gunning, B. E. S. & Robards, A. W. (eds.) (1976). *Intercellular Communication in Plants: Studies on Plasmodesmata*. Berlin: Springer.

Gunning, B. E. S., Hughes, J. E. & Hardham, A. R. (1978). Formative and proliferative cell divisions, cell differentiation, and developmental changes in the meristem of *Azolla* roots. *Planta* 143: 121–44.

Hake, S. & Freeling, M. (1986). Analysis of genetic mosaics shows that the extra epidermal cell divisions in *Knotted* mutant maize plants are induced by adjacent mesophyll cells. *Nature* 320: 621–3.

Hammersley, D. R. H. & McCully, M. E. (1980). Differentiation of wound xylem in pea roots in the presence of colchicine. *Plant Science Letters* 19: 151–6.

Hardham, A. R. & McCully, M. E. (1982). Reprogramming of cells following wounding in pea (*Pisum sativum* L.). I. Cell division and differentiation of new vascular elements. *Protoplasma* 112: 143–51.

Harris, G. P. & Hart, E. M. H. (1964). Regeneration from leaf squares of *Peperomia sandersi* A.D.C.: a relationship between rooting and budding. *Annals of Botany* 28: 509–26.

Haughn, G. W. & Somerville, C. R. (1988). Genetic control of morphogenesis in *Arabidopsis*. *Developmental Genetics* 9: 73–89.

Haupt, W. (1952). Untersuchungen über den Determinationsvorgäng der Blütenbildung bei *Pisum sativum*. *Zeitschrift für Botanik* 40: 1–32.

Haupt, W. (1954). Die Übertragung bluhforderner Prinzipien bei *Pisum sativum* durch Propfung. *Zeitschrift für Botanik* 42: 125–34.

Haupt, W. (1955). Förderung der Blutenbildung durch Hemmung der Vegetativen Entwicklung. *Planta* 46: 403–7.

Hejnowicz, Z. (1955). Growth distribution and cell arrangement in apical meristems. *Acta Societatis Botanicorum Poloniae* 24: 583–608.

Hejnowicz, Z. (1959). Growth and cell division in the apical meristem of wheat roots. *Physiologia Plantarum* 12: 124–38.

Hejnowicz, Z. (1961). Anticlinal divisions, intrusive growth, and loss of fusiform initials in non-storied cambium. *Acta Societatis Botanicorum Poloniae* 30: 729–49.

Hejnowicz, Z. (1967). Interrelationship between mean length, rate of intrusive

elongation, frequency of anticlinal divisions and survival of fusiform initials in cambium. *Acta Societatis Botanicorum Poloniae* 36: 367–78.
Hejnowicz, Z. (1973). Mophogenetic waves in the cambia of trees. *Plant Science Letters* 1: 359–66.
Hejnowicz, Z. (1975). A model for morphogenetic map and clock. *Journal of Theoretical Biology* 54: 345–62.
Hejnowicz, Z. (1980). Tensional stress in the cambium and its developmental significance. *American Journal of Botany* 67: 1–5.
Hejnowicz, Z. & Tomaszewski, M. (1969). Growth regulators in *Pinus sylvestris*. *Physiologia Plantarum* 22: 984–92.
Hepler, P. K. & Fosket, D. E. (1971). The role of microtubules in vessel member differentiation in *Coleus*. *Protoplasma* 72: 213–36.
Hepler, P. K. & Newcomb, E. H. (1963). The fine structure of young tracheary xylem elements arising by redifferentiation of parenchyma in wounded *Coleus* stems. *Journal of Experimental Botany* 14: 496–503.
Hertel, R. & Flory, R. (1968). Auxin movement in corn coleoptiles. *Planta* 82: 123–44.
Hertwig, O. (1896). *The Biological Problem of Today, Preformation or Epigenesis? The Basis of a Theory of Organic Development*. Translated by P. Chalmers Mitchell. New York: Macmillan.
Heslop-Harrison, J. (1963). Sex expression in flowering plants. *Brookhaven Symposia in Biology* 16: 109–25.
Heslop-Harrison, J. (1967). Differentiation. *Annual Review of Plant Physiology* 18: 325–48.
Hess, T. & Sachs, T. (1972). The influence of a mature leaf on xylem differentiation. *The New Phytologist* 71: 903–14.
Hicks, G. S. & Sussex, I. M. (1971). Organ regeneration in sterile culture after median bisection of the flower primordia of *Nicotiana tabacum*. *Botanical Gazette* 132: 350–63.
Hillman, J. R. (1984). Apical dominance. In *Advanced Plant Physiology*, ed. M. B. Wilkins, pp. 127–48. London: Pitman.
Hofmeister, W. (1868). *Allgemeine Morphologie der Gewächse*. Leipzig: Engelmann.
Holder, N. (1979). Positional information and pattern formation in plant morphogenesis and a mechanism for the involvement of plant hormones. *Journal of Theoretical Biology* 77: 195–212.
Holdsworth, M. (1956). The concept of minimum leaf number. *Journal of Experimental Botany* 7: 395–409.
Hooykaas, P. J., Ooms, G. & Schilperoort, R. A. (1982). Tumors induced by different strains of *Agrobacterium tumefaciens*. In *The Molecular Biology of Plant Tumors*, eds. G. Kahl & J. S. Schell, pp. 373–90. New York: Academic Press.
Iannaccone, P. M., Weinberg, W. C. & Berkwits, L. (1987). A probabilistic model of mosaicism based on the histological analysis of chimeric rat liver. *Development* 99: 187–96.
Ishikawa, K., Kamada, H., Yamaguchi, I. Takahashi, N. & Harada, H. (1988). Morphology and hormone levels of tobacco and carrot tissues transformed by *Agrobacterium tumefaciens* I. Auxin and cytokinin contents of cultured tissues transformed with wild-type and mutant Ti plasmids. *Plant and Cell Physiology* 29: 461–6.
Iterson, G. van, Jr. (1907). *Mathematische und mikroskopisch-anatomische Studien über Blattstellungen*. Jena: Fischer-Verlag.

Jackson, J. A. & Lyndon, R. F. (1988). Cytokinin habituation in juvenile and flowering tobacco. *Journal of Plant Physiology* 132: 575–9.
Jackson, M. B. & Cambell, D. J. (1975). Movement of ethylene from roots to shoots, a factor in the responses of tomato plants to waterlogged soil conditions. *The New Phytologist* 74: 397–406.
Jacobs, M. & Gilbert, S. F. (1983). Basal localization of the presumptive auxin carrier in pea stem cells. *Science* 220: 1297–300.
Jacobs, M. & Short, T. W. (1986). Further characterization of the presumptive auxin transport carrier using monclonal antibodies. In *Plant Growth Substances 1985*, ed. M. Bopp, pp. 218–26. Berlin: Springer.
Jacobs, W. P. (1952). The role of auxin in the differentiation of xylem round a wound. *American Journal of Botany* 39: 301–9.
Jacobs, W. P. (1959). What substances normally control a biological process? I. Formulation of some rules. *Developmental Biology* 1: 527–33.
Jacobs, W. P. (1962). The longevity of plant organs: internal factors controlling abscission. *Annual Review of Plant Physiology* 13: 403–36.
Jacobs, W. P. (1970). Regeneration and differentiation of sieve tube elements. *International Review of Cytology* 28: 239–73.
Jacobs, W. P. (1979). *Plant Hormones and Plant Development*. Cambridge: Cambridge University Press.
Jacobs, W. P., Danielson, J., Hurst, V. & Adams, P. (1959). What substances normally control a biological process? II. The relation of auxin to apical dominance. *Developmental Biology* 1: 534–54.
Jaffe, L. F. (1981). The role of ionic currents in establishing developmental pattern. *Philosophical Transactions of the Royal Society of London*, B 295: 553–66.
Jaffe, L. F. & Nuccitelli, R. (1977). Electrical control of development. *Annual Review of Biophysics and Bioenergetics* 6: 445–76.
Janse, J. M. (1914). Les sections annulaires de l'écroce et le suc descendant. *Annales du Jardin Botanique de Buitenzorg* 28: 1–90.
Jean, R. V. (1984). *Mathematical Approach to Pattern and Form in Plant Growth*. New York: John Wiley & Sons.
Jeffree, C. E. & Yeoman, M. M. (1983). Development of intercellular connections between opposing cells in graft union. *The New Phytologist* 93: 491–509.
Johri, M. M. & Coe, E. H., Jr. (1983). Clonal analysis of corn plant development. I. The development of the tassel and ear shoot. *Developmental Biology* 97: 154–72.
Jost, L. (1893). Über Beziehungen zwischen der Blattentwicklung und der Gefässbildung in der Pflanze. *Botanisches Zeitschrift* 51: 89–138.
Jost, L. (1907). *Lectures on Plant Physiology*, translated by R. J. Harvey Gibson. Oxford: Clarendon Press.
Jost, L. (1942). Über Gefässbrücken. *Zeitschrift für Botanik* 38: 161–215.
Jensen, L. C. W. (1971). Experimental bisection of *Aquilegia* floral buds cultured in vitro. I The effect on growth, primordia initiation, and apical regeneration. *Canadian Journal of Botany* 49: 487–93.
Juniper, B. E. (1977). Some speculations on the possible roles of plasmodesmata in the control of differentiation. *Journal of Theoretical Biology* 66: 583–92.
Juniper, B. E. & Barlow, P. W. (1969). The distribution of plasmodesmata in the root tip of maize. *Planta* 89: 352–60.
Kadej, F. R. (1970). Apical meristem regeneration in root of *Raphanus sativus*. *Acta Societatis Botanicorum Poloniae* 39: 373–81.
Kahl, G. & Schell, J. S. (eds.) (1982). *Molecular Biology of Plant Tumors*. New York: Academic Press.

Khaner, O. & Eyal-Giladi, H. (1986). The embryo-forming potency of the posterior marginal zone in stages X through XII of the chick. *Developmental Biology* 115: 275–81.

Kaldewey, H. (1984). Transport and other modes of movement of hormones (mainly auxins). *Encyclopedia of Plant Physiology*, New Series, vol. 10: 80–148.

Kauffman, S. A. (1973). Control circuits for determination and transdetermination. *Science* 181: 310–18.

Keeble, F., Nelson, M. G. & Snow, R. (1930). The integration of plant behaviour. II. The influence of the shoot on the growth of roots in seedlings. *Proceedings of the Royal Society of London*, B 106: 182–8.

Kende, H. (1965). Kinetin-like factors in the root exudate of sunflowers. *Proceedings of the National Academy of Sciences of the U.S.A.* 53: 1302–7.

Kerbauy, G. B., Monteiro, W. R., Kraus, J. E. & Hell, K. G. (1988). Some physiological and structural aspects of cytokinin-autonomy in the callus of tobacco (*Nicotiana tabacum* L.). *Journal of Plant Physiology* 132: 218–22.

Khait, A. (1986). Hormonal mechanisms for size measurement in living organisms in the context of maturing juvenile plants. *Journal of Theoretical Biology* 118: 471–3.

Kinet, J. M., Sachs, R. M. & Bernier, G. (1981). *The Physiology of Flowering*, volume 3. Boca Raton, Florida: C.R.C. Press.

King, P. J. (1988). Plant hormone mutants. *Trends in Genetics* 4: 157–62.

Kirschner, H. & Sachs, T. (1972). Correlative inhibition between strips of vascular tissue. *Israel Journal of Botany* 21: 129–34.

Kirschner, H. & Sachs, T. (1978). Cytoplasmic reorientation: an early stage of vascular differentiation. *Israel Journal of Botany* 27: 131–7.

Kirschner, H., Sachs, T. & Fahn, A. (1971). Secondary xylem reorientation as a special case of vascular tissue differentiation. *Israel Journal of Botany* 20: 184–98.

Kirschner, M. & Mitchison, T. (1986). Beyond self-assembly: from microtubules to morphogenesis. *Cell* 45: 329–42.

Klein, R. M. & Weisel, B. W. (1964). Determinant growth in the morphogenesis of bean hypocotyls. *Bulletin of the Torrey Botanical Club* 91: 217–24.

Klekowski, E. J., Jr. (1988). *Mutation, Developmental Selection and Plant Evolution*. New York: Columbia University Press.

Klekowski, E. J., Jr. & Kazarinova-Fukshansky, N. (1984a). Shoot apical meristems and mutation: fixation of selectively neutral cell genotypes. *American Journal of Botany* 71: 22–7.

Klekowski, E. J., Jr. & Kazarinova-Fukshansky, N. (1984b). Shoot apical meristems and mutation: selective loss of disadvantageous cell genotypes. *American Journal of Botany* 71: 28–34.

Kny, L. (1894). On correlation in the growth of roots and shoots. *Annals of Botany* 8: 265–80.

Kollmann, R. & Glockmann, C. (1985). Studies on graft union. I. Plasmodesmata between cells of plants belonging to plants of different, unrelated taxa. *Protoplasma* 124: 224–35.

Kollmann, R., Yang, S. & Glockmann, C. (1985). Studies on graft union. II. Continuous and half plasmodesmata in different regions of the graft interface. *Protoplasma* 126: 19–29.

Korn, R. W. (1972). Arrangement of stomata on the leaves of *Pelargonium zonale*. *Annals of Botany* 36: 325–33.

Korn, R. W. (1981). A neighboring-inhibition model for stomate patterning. *Developmental Biology* 88: 115–20.

Korn, R. W. & Fredrick, G. W. (1973). Development of D-type stomata in the leaves of *Ilex crenata* var. *convexa*. *Annals of Botany* 37: 647–56.
Kuiper, D. (1988). Growth responses of *Plantago major* L. ssp. *pleiosperma* (Pilger) to changes in mineral supply. *Plant Physiology* 87: 555–7.
Küster, E. (1925). *Pathologische Pflanzenanatomie*, 3rd edn. Jena: Gustav Fischer.
Lacalli, T. C. & Harrison, L. G. (1978). Development of ordered arrays of cell wall pores in desmids: a nucleation model. *Journal of Theoretical Biology* 74: 109–38.
Lachaud, S. (1983). Xylogénèse chez les dicotylédones arborescentes. IV. Influence des bourgeons, de l'acide β-índolyl acétique et de l'acide gibberellique sur la réactivation cambiale et la xylogénèse dans les jeunes tiges de Hêtre. *Canadian Journal of Botany* 61: 1768–74.
Lachaud, S. & Bonnemain, J. L. (1982). Xylogénèse chez les dicotylédones arborescentes. III. Transport de l'auxine et activité cambiale dans les jeunes tiges de Hêtre. *Canadian Journal of Botany* 60: 869–76.
Lachaud, S. & Bonnemain, J. L. (1984). Seasonal variations in the polar transport pathways and retention site of (^{3}H)indole-3-acetic acid in young branches of *Fagus sylvatica* L. *Planta* 161: 207–15.
Lang, A. (1973). Inductive phenomena in plant development. *Brookhaven Symposia in Biology* 25: 129–44.
Lang, J. M., Eisinger, W. R. & Green, P. B. (1982). Effects of ethylene on the orientation of microtubules and cellulose microfibrils of pea epicotyl cells with polylamellate cell walls. *Protoplasma* 110: 5–14.
Larson, P. R. (1975). Development and organization of the primary vascular system in *Populus deltoides* according to phyllotaxy. *American Journal of Botany* 62: 1084–99.
Lerman, M. I. (1978). The biological essence of resting cells in cell populations. *Journal of Theoretical Biology* 73: 615–29.
Leshem, B. & Sachs, T. (1986). 'Vitrified' *Dianthus* – teratomata *in vitro* due to growth factor imbalance. *Annals of Botany* 56: 613–17.
Lewis, J. H. & Wolpert, L. (1976). The principle of non-equivalence in development. *Journal of Theoretical Biology* 62: 479–90.
Libbert, E. (1955). Über mögliche Beziehungen zwischen Korrelationshemmstoff und Blütenbildung. *Die Naturwissenschaften* 42: 610–11.
Lindenmayer, A. (1984). Models for plant tissue development with cell division orientation regulated by prophase bands of microtubules. *Differentiation* 26: 1–10.
Lindsay, D. W., Yeoman, M. M. & Brown, R. (1974). An analysis of the development of graft union in *Lycopersicon esculentum*. *Annals of Botany* 38: 639–46.
Lintilhac, P. M. (1984). Positional controls in meristem development: a caveat and an alternative. In *Positional Controls in Plant Development*, eds. P. W. Barlow & D. J. Carr, pp. 83–105. Cambridge: Cambridge University Press.
Loeb, J. (1924). *Regeneration*. New York: McGraw-Hill.
Loiseau, J.-E. (1969). *La Phyllotaxie*. Paris: Masson et Cie.
Lopriore, G. (1895). Verläufige Mitteilung über die Regeneration Gespalteter Stammspitzen. *Berichte der deutschen botanischen Gesellschaft* 13: 410–14.
Lopriore, G. (1904). Künstlich erzeugte Verbanderung bei *Phaseolus multiflorus*. *Berichte der deutschen botanischen Gesellschaft* 22: 394–6.
Lüttge, U. & Higginbotham, N. (1979). *Transport in Plants*. New York: Springer.
Lyndon, R. F. (1970a). Rates of cell division in the shoot apical meristem of *Pisum*. *Annals of Botany* 34: 1–17.
Lyndon, R. F. (1970b). Planes of cell division and growth in the shoot apex of *Pisum*. *Annals of Botany* 34: 19–28.

Lyndon, R. F. (1971). Growth of the surface and inner parts of the pea shoot apical meristem during leaf initiation. *Annals of Botany* 35: 263–70.

Lyndon, R. F. (1977). Interaction processes in the vegetative development and in the transition to flowering at the shoot apex. *Symposia of the Society for Experimental Biology* 31: 221–50.

Lyndon, R. F. (1979). The cellular basis of apical differentiation: In *Differentiation and the Control of Plant Development – Potential for Chemical Modification*, ed. E. C. George (Monograph 3 of the British Plant Growth Regulator Group), pp. 57–73. Wantage: British Plant Growth Regulator Group.

Lyndon, R. F. (1982). Changes in the polarity of growth during leaf initiation in the pea, *Pisum sativum*. *Annals of Botany* 49: 281–90.

Lyndon, R. F. (1983). The mechanism of leaf initiation. In *The Growth and Functioning of Leaves* eds. J. E. Dale & F. L. Milthorpe, pp. 3–24. Cambridge: Cambridge University Press.

Lyndon, R. F. & Battey, N. H. (1985). The growth of the shoot apical meristem during flower initiation. *Biologia Plantarum* (*Praha*) 27: 339–49.

Malacinski, G. M. (ed.) (1984). *Pattern Formation. A Primer in Developmental Biology*. New York: Macmillan.

Marx, A. & Sachs, T. (1977). The determination of stomata pattern and frequency in *Anagallis Botanical Gazette* 138: 385–92.

Matthyse, A. G. & Scott, T. K. (1984). Functions of hormones at the whole plant level. *Encyclopedia of Plant Physiology*, New Series, 10: 219–43.

Mauseth, J. D. (1978). An investigation of the phylogenetic and ontogenetic variability of shoot apical meristems in the Cactaceae. *American Journal of Botany* 65: 326–33.

Mauseth, J. D. (1981). A morphometric study of the ultrastructure of *Echinocereus engelmanii* (Cactaceae). II. The mature, zonate apical meristem. *American Journal of Botany* 68: 96–100.

McCallum, M. (1905). Regeneration in plants. *Botanical Gazette* 40: 97–263.

McCready, C. C. (1963). Movement of growth regulators in plants. I. The transport of 2,4-dichlorophenoxyacetic acid in segments from the petioles of *Phaseolus vulgaris*. *The New Phytologist* 62: 3–18.

McCully, M. E. & Dale, H. M. (1961). Variations in leaf number in *Hippuris*: a study of whorled phyllotaxis. *Canadian Journal of Botany* 39: 611–25.

McDaniel, C. N. (1978). Determination of growth pattern in axillary buds of *Nicotiana tabacum* L. *Developmental Biology* 66: 250–5.

McDaniel, C. N. (1980). Influence of leaves and roots on meristem development in *Nicotiana tabacum* L. cv. Wisconsin 38. *Planta* 148: 462–7.

McDaniel, C. N. (1984). Shoot meristem development. In *Positional Controls in Plant Development*, eds. P. W. Barlow & D. J. Carr, pp. 319–47. Cambridge: Cambridge University Press.

McDaniel, C. N. & Hsu, F. C. (1976). Position-dependent development of tobacco meristems. *Nature* 259: 564–5.

McDaniel, C. N. & Poethig, R. S. (1988). Cell-lineage patterns in the shoot apical meristem of the germination maize embryo. *Planta* 175: 13–22.

McGaw, B. A., Horgan, R., Heald, J. K., Wullems, G. J. & Schilperoort, R. A. (1988). Mass-spectrometric quantitation of cytokinins in tobacco crown-gall tumours induced by mutated octopine Ti plasmids of *Agrobacterium tumefaciens*. *Planta* 176: 230–4.

McIntyre, G. (1977). The role of nutrition in apical dominance. *Symposia of the Society for Experimental Biology* 31: 25–74.

McIntyre, G. & Damson, E. (1988). Apical dominance in *Phaseolus vulagaris*. The triggering effect of shoot decapitation and of leaf excision on growth of lateral buds. *Physiologia Plantarum* 74: 607–14.
Meicenheimer, R. D. (1981). Changes in *Epilobium* phyllotaxis induced by N-1-naphthylphthalamic acid and α-4-chlorophenoxyisobutyric acid. *American Journal of Botany* 68: 1139–54.
Meinhardt, H. (1976). Morphogenesis of lines and nets. *Differentiation* 6: 117–23.
Meinhardt, H. (1982). *Models of Biological Pattern Formation*. London: Academic Press.
Meinhardt, H. (1984). Models of pattern formation and their application to plant development. In *Positional Controls in Plant Development*, eds. P. W. Barlow & D. J. Carr, pp. 1–32. Cambridge: Cambridge University Press.
Meins, F., Jr. (1982). Habituation of cultured plant cells. In *Molecular Biology of Plant Tumors*, eds. G. Kahl & J. S. Schell, pp. 3–31. New York: Academic Press.
Meins, F., Jr. (1983). Heritable variation in plant cell culture. *Annual Review of Plant Physiology*, 34: 327–46.
Meins, F., Jr. (1986). Phenotypic stability and variation in plants. *Current Topics in Developmental Biology* 20: 373–82.
Meins, F., Jr. & Binns, A. N. (1979). Cell determination in plant development. *Biosciences* 29: 221–5.
Meins, F., Jr. & Binns, A. N. (1982). Rapid reversion of cell-division factor habituated cells in culture. *Differentiation* 23: 10–12.
Meins, F., Jr. & Lutz, J. (1979). Tissue-specific variation in the cytokinin habituation of cultured tobacco cells. *Differentiation* 15: 1–6.
Meins, F., Jr. & Lutz, J. (1980). The induction of cytokinin habituation in primary pith explants of tobacco. *Planta* 149: 402–7.
Meins, F., Jr., Foster, R. & Lutz, J. (1983). Evidence for a mendelian factor controlling the cytokinin requirement of cultured tobacco cells. *Developmental Genetics* 4: 129–41.
Meins, F., Jr. & Wenzler, H. (1986). Stability and the determined state. *Symposia of the Society for Experimental Biology* 40: 155–170.
Michaelson, J. (1987). Cell selection in development. *Biological Reviews of the Cambridge Philosophical Society* 62: 115–39.
Miginiac, E. (1978). Some aspects of regulation of flowering: role of correlative factors in photoperiodic plants *Botanical Magazine, Tokyo*, Special Issue 1, 159–73.
Mignotte, Y., Vallade, J. & Bugnon, F. (1987). Sur la notion de cellule initiale. *Bulletin de la Société Botanique de France* 134, Lettres Botaniques 1987, 275–82.
Millington, W. F. (1966). The tendril of *Parthenocissus inserta*: determination and development. *American Journal of Botany* 53; 74–81.
Mingo-Castel, A. M., Gomez-Campo, C., Tortosa, M. E. & Pelacho, A. M. (1984). Hormonal effects on phyllotaxis of *Euphorbia lathyrus* L. *Botanical Magazine, Tokyo*, 97: 171–8.
Mineyuki, Y., Marc, J. & Palevitz, B. A. (1988). Formation of the oblique spindle in dividing guard mother cells of *Allium*. *Protoplasma* 147: 200–3.
Mitchison, G. J. (1977). Phyllotaxis and the Fibonacci series. *Science* 196: 270–5.
Mitchison, G. J. (1980). A model for vein formation in higher plants. *Proceedings of the Royal Society of London*, B 207: 79–109.
Mitchison, G. J. (1981). The polar transport of auxin and vein patterns in plants. *Philosophical Transactions of the Royal Society of London*, B 295: 461–71.

Mohr, H. (1978). Pattern specification and realization in photomorphogenesis. *Botanical Magazine, Tokyo*, Special Issue 1: 199–217.
Mohr, H. (1988). Control of plant development: signals from without – signals from within. *Botanical Magazine*, 101: 79–101.
Moorby, J. & Wareing, P. F. (1963). Ageing in woody plants. *Annals of Botany* 27: 291–308.
Moore, R. (1982). Graft formation in *Kalanchoë blossfeldiana*. *Journal of Experimental Botany* 33: 533–40.
Moore, R. & Walker, D. B. (1981). Studies of vegetative compatibility in higher plants. *American Journal of Botany* 68: 820–30.
Morgan, P. W. & Gausman, H. W. (1966). Effect of ethylene on auxin transport. *Plant Physiology* 41: 45–52.
Morris, D. A. & Arthur, E. D. (1987). Auxin-induced assimilate translocation in bean stem (*Phaseolus vulgaris* L.). *Plant Growth Regulation* 5: 169–81.
Morris, D. A., Kadir, G. O. & Barry, A. J. (1973). Auxin transport in intact pea seedlings (*Pisum sativum* L.): the inhibition of transport by 2, 3, 5-triiodobenzoic acid. *Planta* 110: 173–82.
Morris, D. A. & Thomas, A. G. (1978). A microautoradiographic study of auxin transport in the stem of intact pea seedlings (*Pisum sativum* L.). *Journal of Experimental Botany* 29: 147–57.
Morris, D. A. & Winfield, P. J. (1972). Kinetin transport to axillary buds of dwarf pea (*Pisum sativum* L.). *Journal of Experimental Botany* 23: 346–54.
Mott, R. L. & Cure, W. W. (1978). Anatomy of maize tissue cultures. *Physiologia Plantarum* 42: 139–45.
Mullins, M. G., Nair, Y. & Sampet, P. (1979). Rejuvenation *in vitro*: induction of juvenile characters in an adult clone of *Vitis vinifera* L. *Annals of Botany* 44: 623–7.
Murashige, T. (1974). Plant propagation through tissue culture. *Annual Review of Plant Physiology* **25**: 135–66.
Neeff, F. (1914). Über Zellumlagerung: ein Beitrag zur experimentellen Anatomie. *Zeitschrift für Botanik* 6: 465–547.
Neilson-Jones, W. (1969). *Plant Chimeras*. 2nd edn. London: Methuen.
Němec, B. (1905). Über Regenerationserscheinengen am amgeschnitten Wurzelspitzen. *Berichte der deutschen botanischen Gesellschaft* 23: 113–20.
Němec, B. (1966). Die Frequenz der Zellteilung in der Wurzelspitze von *Phaseolus vulgaris nanus*. *Biologia Plantarum (Praha)* 8: 5–9.
Newman, I. Y. (1956). Pattern in meristems of vascular plants. I. Cell partition in living apices and in the cambial zone in relation to the concepts of initial cells and apical cells. *Phytomorphology* 6: 1–19.
Noël, C. (1946). Recherches anatomiques sur le "crown gall". *Annales des Sciences Naturelles, Botanique et Biologie Végétale*, 11ème series, 7: 87–145.
Northcote, D. H. (1969). The synthesis and metabolic control of polysaccharides and lignin during differentiation of plant cells. *Essays in Biochemistry* 5: 89–137.
Nozeran, R., Bancilhon, L. & Neville, P. (1971). Intervention of internal correlations in the morphogenesis of higher plants. *Advances in Morphogenesis* 9: 1–66.
Olesen, P. (1979). The neck constriction in plasmodesmata. Evidence for a peripheral sphincter-like structure revealed by fixation with tannic acid. *Planta* 144: 349–58.
Osborne, D. J. & McMannus, M. T. (1986). Flexibility and commitment in plant cells during development. *Current Topics in Developmental Biology* 20: 383–96.

Osborne, D. J. & Mullins, M. G. (1969). Auxin, ethylene and kinetin in a carrier – protein model system for the polar transport of auxin in petiole segments of *Phaseolus vulgaris*. *The New Phytologist* 68: 977–91.

Overall, R. L. & Gunning, B. E. S. (1982). Intercellular communication in *Azolla* roots. II. Electrical coupling. *Protoplasma* 111: 151–60.

Overall, R. L., Wolfe, J. & Gunning, B. E. S. (1982). Intercellular communication in *Azolla* roots. I. Ultrastructure of plasmodesmata. *Planta* 41: 134–50.

Overbeek, J. van (1938). Auxin distribution in seedlings and its bearing on the problem of apical dominance. *Botanical Gazette* 100: 133–66.

Palevitz, B. A. & Helper, P. K. (1985). Changes in dye coupling of stomatal cells of *Allium* and *Commelina* demonstrated by microinjection of Lucifer Yellow. *Planta* 164: 473–9.

Palni, L. M. S., Burch, L. & Horgan, R. (1988). The effect of auxin concentration on cytokinin stability and metabolism. *Planta* 174: 231–4.

Pant, D. D. & Kidwai, P. F. (1967). Development of stomata in some Cruciferae. *Annals of Botany* 31: 513–21.

Parkinson, M. & Yeoman, M. M. (1982). Graft formation in cultured, explanted internodes. *The New Phytologist* 91: 711–9.

Parkinson, M., Jeffree, C. E. & Yeoman, M. M. (1987). Incompatibility in cultured explant grafts between members of the Solanaceae. *The New Phytologist* 107: 489–98.

Paton, D. M. (1978). Node of flowering as an index of plant development: a further examination. *Annals of Botany* 42: 1007–8.

Paton, D. M. & Barber, H. N. (1955). Physiological genetics of *Pisum*. T. Grafting experiments between early and late varieties. *Australian Journal of Biological Sciences* 8: 231–40.

Patrick, J. W. (1976). Hormone directed transport of metabolites. In *Transport and Transfer Processes of Plants*, eds. I. F. Wardlaw & J. B. Passioura, pp. 433–46. New York: Academic Press.

Patrick, J. W. & Wareing, P. F. (1976). Auxin promoted transport of metabolites in stems of *Phaseolus vulgaris* L. Effects at the site of hormone application. *Journal of Experimental Botany* 27: 969–82.

Patrick, J. W. & Woolley, D. J. (1973). Auxin physiology of decapitated stems of *Phaseolus vulgaris* L., treated with indol-3yl-acetic acid. *Journal of Experimental Botany* 24: 949–57.

Pellegrini, O. (1961). Modificazione delle prospettive morfogenetiche in primordi fogliari chirurgicamente isolati dal meristema apicale del germoglio. *Delpinoa*, New Series, 3: 1–12.

Pengelly, W. L. & Meins, F., Jr. (1983). Growth, auxin requirement, and indole-3-acetic acid content of cultured crown-gall and habituated tissues of tobacco. *Differentiation* 25: 101–5.

Philipson, W. R., Ward, J. M. & Butterfield, B. G. (1971). *The Vascular Cambium. Its Development and Activity*. London: Chapman and Hall.

Phillips, I. D. J. (1975). Apical dominance. *Annual Review of Plant Physiology* 26: 341–67.

Phillips, R. (1980). Cytodifferentiation. *International Review of Cytology* 11A: 55–70.

Pickard, B. G. (1973). Action potentials in higher plants. *The Botanical Review* 39: 172–201.

Pilkington, M. (1929). The regeneration of the stem apex. *The New Phytologist* 28: 37–53.

Plantefol, L. (1948). *La Théorie des Hélices Foliares Multiples*. Paris: Massonnet Cie.
Poethig, R. S. (1987). Clonal analysis of cell lineage patterns in plant development. *American Journal of Botany* **74**: 581–94.
Poethig, R. S. (1988). A non-cell-autonomous mutation regulating juvenility in maize. *Nature* 336: 82–3.
Poething, R. S. & Sussex, I. M. (1985a). The developmental morphology and growth dynamics of the tobacco leaf. *Planta* 165: 158–69.
Poething, R. S. & Sussex, I. M. (1985b). The cellular parameters of leaf development in tobacco: a clonal analysis. *Planta* 165: 170–84.
Prantl, K. (1874). Untersuchungen über die Regeneration des Vegetationspunktes am Angiospermenwurzeln. *Arbeiten aus dem botanisches Institut Würzburg* 1: 546–62.
Priestley, J. H. & Swingle, C. F. (1929). Vegetative propragation from the standpoint of plant anatomy. *United States Department of Agriculture Technical Bulletin*, no. 151: 1–98.
Prigogine, I. & Stengler, I. (1984). *Order Out of Chaos*. London: W. Heinemann.
Pruitt, R. E., Chang, C., Pang, P. P-Y. & Meyerowitz, E. M. (1987). Molecular genetics and development in *Arabidopsis. In Genetic Regulation of Development*, ed. W. F. Loomis, pp. 327–38. New York: Alan Liss.
Raff, J. W., Hutchinson, J. F., Knox, R. B. & Clarke, A. E. (1979). Cell recognition: antigenic determinants of plant organs and their cultured cells. *Differentiation* 12: 179–86.
Raju, M. V. S., Marchuk, W. N. & Polowick, P. L. (1978). Surgical studies on the growth and xylem differentiation in cotyledonary shoots of flax. *Canadian Journal of Botany* 56: 476–82.
Raschke, K. (1975). Stomata action. *Annual Review of Plant Physiology* 26: 309–40.
Rasmussen, H. (1981). Terminology and classification of stomata and stomata development – a critical survey. *Botanical Journal of the Linnean Society* 83: 199–212.
Rasmussen, H. (1986a). Epidermal cell differentiation during leaf development in *Anemarrhena asphodeloides* (Liliaceae). *Canadian Journal of Botany* 64: 1277–85.
Rasmussen, H. (1986b). Pattern formation and cell interactions in epidermal development of *Anemarrhena asphodeloides* (Liliaceae). *Nordic Journal of Botany* 6: 467–77.
Rathore, K. S., Hodges, T. K. & Robinson, K. R. (1988). Ionic basis of currents in somatic embryos of *Daucus carota*. *Planta* 175: 280–9.
Rayle, D. L, Ouitrakul, R. & Hertel, R. (1969). Effects of auxins on the auxin transport system in coleoptiles. *Planta* 87: 49–53.
Reid, J. B. & Murfet, I. C. (1984). Flowering in *Pisum*: a fifth locus, *veg*. *Annals of Botany* 53: 369–82.
Reinert, J. (1982). *Plant Cell and Tissue Culture. A Laboratory Manual*. Berlin: Springer.
Reinhard, E. (1953). Beobachtungen an *in vitro* kultivierten Geweben aus dem Vegetationskegel der *Pisum*-Wurzel. *Zeitschrift für Botanik* 42: 353–76.
Richards, F. J. (1951). Phyllotaxis: its quantitative expression and relation to growth in the apex. *Philosophical Transactions of the Royal Society of London*, B 235; 509–64.
Richmond, A. E. & Lang, A. (1957). Effect of kinetin on protein content and survival of detached *Xanthium* leaves. *Science* 125: 650–1.
Rijven, A. H. G. C. (1968). Randomness in the genesis of phyllotaxis. I. The initiation of the first leaf in some Trifoliae. *The New Phytologist* 67: 247–56.

Robbins, W. J. (1960). Further observations on juvenile and adult *Hedera*. *American Journal Botany* 47: 485–91.
Roberts, I. N., Lloyd, C. W. & Roberts, K. (1985). Ethylene-induced microtubule reorientations: mediation by helical arrays. *Planta* 164: 439–47.
Roberts, L. W. (1960). Experiments on xylem regeneration in stem wound responses in *Coleus*. *Botanical Gazette* 121: 201–8.
Roberts, L. W. (1976). *Cytodifferentiation in Plants*. Cambridge: Cambridge University Press.
Robinson, L. W. & Wareing, P. F. (1969). Experiments on the juvenile – adult phase change in some woody species. *The New Phytologist* 68: 67–78.
Rogler, C. E. & Hackett, W. P. (1975). Phase change in *Hedera helix*: induction of mature and juvenile phase changes by gibberellin A_3. *Physiologia Plantarum* 34: 141–7.
Rost, T. L. & Jones, T. J. (1988). Pea root regeneration after tip excisions at different levels: polarity of new growth. *Annals of Botany* 61: 513–23.
Rubery, P. H. (1987). Auxin transport. In *Plant Hormones and their Role in Plant Growth and Development*, ed. P. J. Davies, pp. 341–62. Dordrecht: Martinus Nijhoff.
Rubinstein, B. & Nagao, M. A. (1976). Lateral bud growth and its control by the apex. *The Botanical Review* **42**: 83–113.
Ruge, U. (1937). Untersuchungen über den Einfluss des Hetero-auxins auf das Streckungswachstum des Hypokotyls von *Helianthus annuus*. *Zeitschrift für Botanik* 31: 1–56.
Rutishauser, R. (1981). *Blattstellung und Sprossentwicklung bei Blütenpflanzen*. Vaduz: Cramer.
Rutishauser, R. & Sattler, R. (1986). Architecture and development of the phyllode-stipule whorls of *Acacia longipedunculata*: controversial interpretation and continuum approach. *Canadian Journal of Botany* 64: 1987–2019.
Rutishauser, R. & Sattler, R. (1987). Complementary and heuristic value of contrasting models in structural botany. II. Case study on leaf whorls: *Equisetum* and *Ceratophyllum*. *Botanische Jahrbücher für Systematik* 109: 227–55.
Ryleski, I. & Halevy, A. H. (1972). Factors controlling the readiness to flower of various buds along the main axis. *Journal of the American Society in Horticultural Sciences* 97: 309–12.
Sachs, T. (1966). Senescence of inhibited shoots of peas and apical dominance. *Annals of Botany* 30: 447–56.
Sachs, T. (1968). The role of the root in the induction of xylem differentiation in peas. *Annals of Botany* 32: 391–9.
Sachs, T. (1969a). Polarity and the induction of organized vascular tissues. *Annals of Botany* 33: 263–75.
Sachs, T. (1969b). Regeneration experiments on the determination of the form of leaves. *Israel Journal of Botany* 18: 21–30.
Sachs, T. (1970). A control of bud growth by vascular tissue differentiation. *Israel Journal of Botany* 19: 484–98.
Sachs, T. (1972a). The induction of fibre differentiation in peas. *Annals of Botany* 36: 189–97.
Sachs, T. (1972b). A possible basis for apical organization in plants. *Journal of Theoretical Biology* 37: 353–61.
Sachs, T. (1974). The developmental origin of stomata pattern in *Crinum*. *Botanical Gazette* 135: 197–204.
Sachs, T. (1975a). The control of the differentiation of vascular networks. *Annals of Botany* 39, 197–204.

Sachs, T. (1975b). Plant tumors resulting from unregulated hormone synthesis. *Journal of Theoretical Biology* 55: 445–53.
Sachs, T. (1978a). Patterned differentiation in plants. *Differentiation* 11: 65–73.
Sachs, T. (1978b). The development of the spacing pattern in the leaf epidermis. In *The Clonal Basis of Development*, eds S. Subtelny & I. M. Sussex, pp. 161–83. New York: Academic Press.
Sachs, T. (1979). Cellular interactions in the development of stomatal pattern in *Vinca major*. *Annals of Botany* 43: 693–700.
Sachs, T. (1981a). The controls of the patterned differentiation of vascular tissues. *Advances in Botanical Research* 9: 151–262.
Sachs, T. (1981b). Polarity changes and tissue organization in plants. In *Cell Biology, 1980–1981*, ed. H. G. Schweiger, pp. 489–96. Berlin: Springer Verlag.
Sachs, T. (1982). A morphogenetic basis for plant morphology. *Acta Biotheoretica* 31A: 118–31.
Sachs, T. (1983). Signal flow as a basis for organized differentiation. In *Biological Structures and Coupled Flows*, eds. A. Oplatka & M. Balban, pp. 457–71. New York: Academic Press.
Sachs, T. (1984a). Axiality and polarity in vascular plants. In *Positional Controls in Plant Development*, eds. P. W. Barlow & D. J. Carr, pp. 193–224. Cambridge: Cambridge University Press.
Sachs, T. (1984b). Controls of cell patterns in plants. In *Pattern Formation*, eds. G. M. Malacinski & S. V. Bryant, pp. 367–91. New York: Macmillan.
Sachs, T. (1986). Cellular interactions in tissue and organ development. *Symposia of the Society for Experimental Biology* 40: 181–210.
Sachs, T. (1988a). Ontogeny and phylogeny: phytohormones as indicators of labile changes. In *Plant Evolutionary Biology*, eds. L. D. Gottlieb & S. K. Jain, pp. 157–76. London: Chapman & Hall.
Sachs, T. (1988b). Epigenetic selection: an alternative mechanism of pattern formation. *Journal of Theoretical Biology* 134: 547–59.
Sachs, T. (1988c). Internal controls of plant morphogenesis. In: *Proceedings of the XIVth International Botanical Congress, Berlin*, eds. W. Greuter & B. Zimmer, pp. 241–60. Königstein: Koeltz.
Sachs, T. & Benouaiche, P. (1978). A control of stomata maturation in *Aeonium*. *Israel Journal of Botany* 27: 47–53.
Sachs, T. & Cohen, D. (1982). Circular vessels and the control of vascular differentiation in plants. *Differentiation* 21: 22–6.
Sachs, T. & Thimann, K. V. (1967). The role of auxins and cytokinins in the release of buds from dominance. *American Journal of Botany* **54**: 136–44.
Sagromsky, H. (1949). Weitere Beobachtungen zur Bildung der Spaltöffnungsmusters in der Blattepidermis. Zur Frage der Gruppenbildung. *Zeitschrift für Naturforschung* 4B: 360–7.
Sakaguchi, S., Hogetsu, T. & Hara, N. (1988). Arrangement of cortical microtubules at the surface of the shoot apex in *Vinca major* L.: observations by immunofluorescence microscopy. *Botanical Magazine, Tokyo* 101: 497–507.
Sander, K. & Nübler-Jung, K. (1981). Polarity and gradients in insect development. In *Cell Biology, 1980–1981*, ed. H. G. Schweiger, pp. 497–511. Berlin: Springer-Verlag.
Santiago, J. F. & Goodwin, P. B. (1988). Restricted cell/cell communication in the shoot apex of *Silene coeli-rosa* during the transition to flowering is associated with a high mitotic index rather than with evocation. *Protoplasma* 146: 52–60.
Sass, J. E. (1932). The formation of callus knots as related to the histology of graft union. *Botanical Gazette* 94: 364–80.

Satina, S., Blakeslee, A. F. & Avery, A. G. (1940). Demonstration of the three germ layers in the shoot apex for *Datura* by means of induced polyploidy in periclinical chimeras. *American Journal of Botany* 27: 895–905.
Sattler, R. (1973). *Organogenesis of Flowers: A Photographic Text-Atlas.* Toronto: University of Toronto Press.
Sattler, R. (1966). Towards a more adequate approach to comparative morphology. *Phytomorphology* 165: 417–29.
Sattler, R. (1974). A new conception of the shoot in higher plants. *Journal of Theoretical Biology* 47: 367–82.
Savidge, R. A. & Wareing, P. F. (1981). A tracheid differentiation factor from pine needles. *Planta* 153: 395–404.
Sax, K. & Dickson, A. Q. (1956). Phloem polarity in bark regeneration. *Journal of the Arnold Arboretum* 37: 173–9.
Schaffalitky de Muckadell, M. (1954). Juvenile stages in woody plants. *Physiologia Plantarum* 7: 782–96.
Schiavone, F. M. (1988). Microamputation of somatic embryos of the domestic carrot reveals apical control of axis elongation and root regeneration. *Development* 103: 657–64.
Schnall, J. A., Cooke, T. J. & Cress, D. E. (1988). Genetic analysis of somatic embryogenesis in carrot cell culture: initial characterization of six classes of temperature sensitive variants. *Developmental Genetics* 9: 49–67.
Schwabe, W. W. (1984). Phyllotaxis. In *Positional Controls in Plant Development*, eds. P. W. Barlow & D. J. Carr, pp. 403–40. Cambridge: Cambridge University Press.
Schwabe, W. W. & Al-Doori, A. H. (1973). Analysis of juvenile-like condition affecting flowering in the black current (*Ribes nigrum*). *Journal of Experimental Botany* 24: 969–81.
Schnepf, E. (1986). Cellular polarity *Annual Review of Plant Physiology* 37: 23–47.
Schoch, P-G., Zinsou, C. & Sibu, M. (1980). Dependence of stomatal index on environmental factors during stomata differentiation in leaves of *Vigna sinensis* L. *Journal of Experimental Botany* 31: 1211–16.
Schulz, A. (1986a). Wound phloem in transition to bundle phloem in primary roots of *Pisum sativum* L. I. Development of the bundle-leaving wound-sieve tubes. *Protoplasma* 130: 12–26.
Schulz, A. (1986b). Wound phloem in transition to bundle phloem in primary roots of *Pisum sativum* L. II. The plasmatic contact between wound sieve tubes and regular phloem. *Protoplasma* 130: 27–40.
Schulz, A. (1988). Vascular differentiation in the root cortex of peas: premitotic stages of cytoplasmic reactivation. *Protoplasma* 143: 176–87.
Scott, T. K. & Briggs, W. R. (1960). Auxin relationships in the Alaska pea (*Pisum sativum*). *American Journal of Botany* 47: 492–9.
Sheldrake, A. R. (1971). Auxin in the cambium and its differentiating derivatives. *Journal of Experimental Botany* 22: 735–40.
Sheldrake, A. R. (1973). Auxin transport in secondary tissues. *Journal of Experimental Botany* 24: 87–96.
Sheldrake, A. R. (1974). The polarity of auxin transport in inverted cuttings. *The New Phytologist* **73**: 637–42.
Sheldrake, A. R. & Northcote, D. H. (1968). The production of auxin by tobacco internode tissues. *The New Phytologist* 67: 1–13.
Simon, S. (1908). Experimentelle Untersuchungen über die Entstehung von Gefässverbindungen. *Berichte der deutschen botanischen Gesellschaft*, 26 (Festschrift), 364–96.
Sinnott, E. W. (1960). *Plant Morphogenesis.* New York: McGraw-Hill.

Sinnott, E. W. (1963). *The Problem of Organic Form.* New Haven: Yale University Press.
Sinnott, E. W. & Bloch, R. (1939). Changes in intercellular relations during the growth and differentiation of living plant tissues. *American Journal of Botany* 26: 625–34.
Sinnott, E. W. & Bloch, R. (1945). The cytoplasmic basis of intercellular patterns in vascular differentiation. *American Journal of Botany* 32: 151–6.
Sinnott, E. W. & Bloch, R. (1946). Comparative studies in the air roots of *Monstera deliciosa. American Journal of Botany* 33: 587–90.
Skok, J. (1968). Morphogenetic responses of debudded tobacco plants to gibberellic acid and indoleacetic acid. *Plant Physiology* **43**: 215–23.
Skoog, F. & Miller, C. O. (1957). Chemical regulation of growth and organ formation in plant tissues cultured in vitro. *Symposium of the Society for Experimental Biology* 11: 118–31.
Smith, G. E. (1935). On the orientation of stomata. *Annals of Botany* 49: 451–77.
Smith, T. A. (1985). Polyamines. *Annual Review of Plant Physiology* 36: 117–43.
Snow, M. & Snow, R. (1931). Experiments on phyllotaxis. I. The effect of isolating a primordium. *Philosophical Transactions of the Royal Society of London*, B 221: 1–43.
Snow, M. & Snow, R. (1933). Experiments on phyllotaxis. II. The effect of displacing a primordium. *Philosophical Transactions of the Royal Society of London*, B 222: 353–400.
Snow, M. & Snow, R. (1935). Experimental on phyllotaxis. III. Diagonal splits through deccusate apices. *Philosophical Transactions of the Royal Society of London*, B. 225: 63–94.
Snow, M. & Snow, R. (1947). On the determination of leaves. *The New Phytologist* 46: 5–19.
Snow, M. & Snow, R. (1951). On the question of tissue tensions in stem apices. *The New Phytologist* 50: 184–5.
Snow, M. & Snow, R. (1952). Minimum areas and leaf determination. *Proceedings of the Royal Society of London*, B 139: 545–66.
Snow, M. & Snow, R. (1955). Regulation of sizes of leaf primordia by the growing point of stem apex. *Proceedings of the Royal Society of London*, B 144: 222–8.
Snow, M. & Snow, R. (1959). The dorsiventrality of leaf primordia. *The New Phytologist* 58: 188–207.
Snow, M. & Snow, R. (1962). A theory of the regulation of phyllotaxis based on *Lupinus albus. Philosophical Transactions of the Royal Society of London*, B 244: 483–514.
Snow, R. (1929). The young leaf as the inhibitory organ. *The New Phytologist* 28: 345–58.
Snow, R. (1931). Experiments on growth and inhibition. II. New phenomena of inhibition. *Proceedings of the Royal Society of London*, B 108: 305–16.
Snow, R. (1938). On the upwards inhibiting effect of auxin in shoots. *The New Phytologist* 37: 173–85.
Snow, R. (1942). Further experiments on whorled phyllotaxis. *The New Phytologist* 41: 108–24.
Snow, R. (1945). Plagiotropism and correlative inhibition. *The New Phytologist* 44: 110–17.
Snow, R. (1952). On the shoot apex and phyllotaxis of *Costus. The New Phytologist* 51: 359–63.
Snow, R. (1955). Problems of phyllotaxis and leaf determination. *Endeavour* 14: 190–9.

Snow, R. (1965). The causes of bud eccentricity and the large divergence angles between leaves in Cucurbitaceae. *Philosophical Transactions of the Royal Society*, B 250: 53–77.

Söding, H. (1952). *Die Wuchsstofflehre*. Stuttgart: Thieme.

Soetiarto, S. R. & Ball, E. (1969). Ontogenetical and experimental studies of the floral apex of *Portulaca grandiflora*. 2. Bisection of the meristem in successive stages. *Canadian Journal of Botany* 47: 1067–76.

Soma, K. (1958). Morphogenesis in the shoot apex of *Euphorbia lathyrus* L. *Journal of the Faculty of Science of Tokyo University* III. 7: 199–256.

Soma, K. & Kuriyama, K. (1970). Phyllotactic change in the shoot apex of *Ambrosia artemisiafolia* var. *elatior* during ontogenesis. *Botanical Magazine (Tokyo)* 83: 13–20.

Sossountzov, L., Maldiney, R., Sotta, B., Sabbagh, I., Habricot, Y., Bonnet, M. & Miginiac, E. (1988). Immunocytochemical localization of cytokinins in Craigella tomato and a sideshootless mutant. *Planta* 175: 291–304.

Spemann, H. (1938). *Embryonic Development and Induction*. New Haven: Yale University Press.

Sprent, J. I. (1966). Role of the leaf in flowering of late pea varieties. *Nature* 209: 1043–4.

Stebbins G. L. & Jain, S. J. (1960). Developmental studies of cell differentiation in the epidermis of Monocotyledons. *Developmental Biology* 2: 409–26.

Stebbins, G. L. & Shah, S. S. (1960). Developmental studies of cell differentiation in the epidermis of Monocotyledons. II. Cytological features of stomatal development in the Gramineae. *Developmental Biology* 2: 477–500.

Steeves, T. A. & Sussex, I. M. (1972). *Patterns in Plant Development* Englewood Cliffs, New Jersey: Prentice Hall.

Stein, O. L. & Steffensen, D. M. (1959). Radiation–induced markers in the study of leaf growth in *Zea*. *American Journal of Botany* 46: 485--9.

Stern, C. (1968). *Genetic Mosaics and Other Essays*. Cambridge, Mass.: Harvard University Press.

Steward, F. C. (1968). *Growth and Organization in Plants*. Reading, Mass.: Addison-Wesley.

Stewart, R. N. & Dermen, H. (1975). Flexibility in ontogeny as shown by the contribution of the shoot apical layers to leaves of periclinal chimeras. *American Journal of Botany* 62: 935–47.

Stieber, J. (1985). Wave nature and a theory of cambial activity. *Canadian Journal of Botany* 63: 1942–50.

Stoddard, F. L. & McCully, M. E. (1980). Effects of excision of stock and scion organs on the formation of the graft union in *Coleus*. A histological study. *Botanical Gazette* 141: 401–12.

Stoutemyer, V. T. & Britt, O. K. (1965). The behaviour of tissue cultures from English and Algerian ivy in different growth phases. *American Journal of Botany* 52: 805–10.

Sulston, J. E., Schierenberg, E., White, J. G. & Thomson, J. N. (1983). The embryonic cell lineage of the nematode *Caenorhabditis elegans*. *Developmental Biology* 100: 64–119.

Sussex, I. M. (1952). Regeneration of the potato shoot apex. *Nature* 170: 755–7.

Sussex, I. M. (1955). Experimental investigation of leaf dorsiventrality and orientation in the juvenile shoot. *Phytomorphology* 5: 286–300.

Sussex, I. M. & Steeves, T. A. (1967). Apical initials and the concept of the promeristem. *Phytomorphology* 17: 387–91.

Syono, K. & Furuya, T. (1972). Effects of cytokinins on the auxin requirement and auxin contect of tobacco calluses. *Plant and Cell Physiology* 13: 843–56.

Tepper, H. B. & Hollis, C. A. (1967). Mitotic reactivation of the terminal bud and cambium of white ash. *Science* 156: 1635–6.

Thair, B. W. & Steeves, T. A. (1976). Response of the vascular cambium to reorientation in patch grafts. *Canadian Journal of Botany* 54: 361–73.

Thimann, K. V. (1937). On the nature of inhibitions caused by auxin. *American Journal of Botany* 24: 407–12.

Thimann, K. V. (1977). *Hormone Action in the Life of the Whole Plant.* Amherst: University of Massachusetts Press.

Thimann, K. V., Sachs, T. & Mathur, K. N. (1971). The mechanism of apical dominance in *Coleus. Physiologia Plantarum* 24: 68–72.

Thimann, K. V. & Skoog, F. (1933). Studies on the growth hormone of plants. III. The inhibiting action of growth substance on plant development. *Proceedings of the National Academy of Sciences of the U.S.A.* 19: 714–16.

Thimann, K. V. & Skoog, F. (1934). On the inhibition of development and other functions of growth substances in *Vicia faba. Proceeding of the Royal Society of London* B 114: 317–39.

Thomas, B. & Vince-Prue, D. (1984). Juvenility, photoperiodism and vernalization. In *Advanced Plant Physiology* ed. M. B. Wilkins, pp. 408–39. London: Pitman.

Thomson, B. F. & Miller, P. M. (1950). The role of light in histogenesis and differentiation in the shoot of *Pisum sativum.* I. The apical region. *American Journal of Botany* 49: 303–10.

Tilney-Bassett, R. A. E. (1963). The structure of the periclinal chimeras. I. The analysis of periclinal chimeras. *Heredity* 18: 265–85.

Tilney-Bassett, R. A. E. (1986). *Plant Chimeras.* London: Edward Arnold.

Torrey, J. G. (1950). The induction of lateral roots by indoleactic acid and by root decapitation. *American Journal of Botany* 37: 257–64.

Torrey, J. G. (1955). On the determination of vascular patterns during tissue differentiation in excised pea roots. *American Journal of Botany* 42: 183–98.

Torrey, J. G. (1957). Auxin control of vascular pattern formation in regenerating pea root meristems growth *in vitro. American Journal of Botany* 44: 859–70.

Tran Thanh Van, M. (1973). Direct flower neoformation from superficial tissue of small explants of *Nicotiana tabacum* L. *Planta* 115: 87–92.

Trewavas, A. (1981). How do plant growth substances work? *Plant, Cell and Environment* 4: 203–8.

Trewavas, A. J. (1987). Many a mickle makes a muckle; network maxims may mitigate controversy. *BioEssays* 7: 84–6.

Tucker, S. C. & Hoefert, A. A. (1968). Ontogeny of the tendril of *Vitis vinifera. American Journal of Botany* 55: 1110–19.

Turgeon, R. (1982). Teratomas and secondary tumors. In *Molecular Biology of Plant Tumors,* eds. G. Kahl & J. S. Schell, pp. 391–414. New York: Academic Press.

Turing, A. M. (1952). The chemical basis of morphogenesis. *Philosophical Transactions Royal Society of London* B 237: 37–72.

Umrath, K. (1948). Dornenbildung, Blattform und Blütenbildung in abhängigkeit von Wuchsstoff und korrelativer Hemmung. *Planta* 36: 262–97.

Vasil, I. K. (ed.) (1981). Perspectives in Plant Cell and Tissue Culture. *International Review of Cytology, Supplements IIa and IIb.*

Venverloo, C. J. (1976). The formation of adventitious organs. III. A comparison of root and shoot formation on *Nautilocalyx* explants. *Zeitschrift für Pflanzenphysiologie* 80: 310–22.

Venverloo, C. J., Koster, J. & Libbenga, K. R. (1983). The formation of adventitious organs. IV. The ontogeny of shoots and leaves from epidermis cells of *Nautilocalyx* x *lynchii*. *Zeitschrift für Pflanzenphysiologie* 109: 55–67.

Vöchting, H. (1878). *Über Organbildung im Pflanzenreich*, vol. 1. Bonn: Max Cohen.

Vöchting, H. (1892). *Über Transplantation am Pflanzenkörper*. Tubingen: Verlag H. Laupp'schen Buchhandlung.

Vöchting, H. (1904). Über die Regeneration der *Araucaria excelsa*. *Jahrbücher für wissenschaftliche Botanik* 40: 144–55.

Vöchting, H. (1906). Über Regeneration und Polarität bei höher Pflanzen. *Botanisches Zeitung* 64: 101–48.

Waddington, C. H. (1966). Fields and gradients. In *Major Problems in Developmental Biology*, ed. M. Locke, pp. 105–24. New York: Academic Press.

Walker, D. B. (1975). Postgenial fusion in *Catharanthus roseus*. III. Fine structure of the epidermis during and after fusion. *Protoplasma* 86: 29–41.

Walker, D. B. & Bruck, D. K. (1985). Incompetence of stem epidermal cells to dedifferentiate and graft. *Canadian Journal of Botany* 63: 2129–32.

Wakhloo, J. L. (1970). Role of mineral nutrients and growth regulators in the apical dominance in *Solanum sisymbrifolium*. *Planta* 91: 190–4.

Wang, T. L. & Wareing, P. F. (1979). Cytokinins and apical dominance in *Solanum andigena*: lateral shoot growth and endogenous cytokinin levels in the absence of roots. *The New Phytologist* 82: 19–28.

Wangermann, E. (1967). The effect of the leaf on the differentiation of primary xylem in the internode of *Coleus blumei* Benth. *The New Phytologist* 66: 747–54.

Wangermann, E. (1977). Further localization of auxin transport through internode segments. *The New Phytologist* 79: 501–4.

Wardell, W. L. & Skoog, F. (1969). Flower formation in excised tobacco stem segments. *Plant Physiology* 44: 1402–6.

Wardlaw, C. W. (1946). Experimental and analytical studies of pteridophytes. VII. Stelar morphology: the effect of defoliation on the stele of *Osmunda* and *Todea*. *Annals of Botany* 10: 97–107.

Wardlaw, C. W. (1950). Experimental and analytical studies of pteridophytes. XVI. The induction of leaves and buds in *Dryopteris aristata* Druce. *Annals of Botany* 14: 435–55.

Wardlaw, C. W. (1952). The effect of isolating the apical meristem in *Echinops*, *Nupha*, *Gunnera* and *Phaseolus*. *Phytomorphology* 2: 240–2.

Wardlaw, C. W. (1956). Further investigations on the effect of undercutting fern leaf primordia. *Annals of Botany* 20: 121–32.

Wardlaw, C. W. (1968). *Morphogenesis in Plants*. London: Methuen.

Wardlaw, C. W. & Mitra, G. C. (1957). Responses of the fern apex to gibberellic acid, kinetin and naphthalene acetic acid. *Nature* 181: 400–1.

Wareing, P. F. (1959). Problems of juvenility and flowering in trees. *Journal of the Linnean Society, Botany* 56: 282–9.

Wareing, P. F. (1978). Determination in plant development. *Botanical Magazine, Tokyo*, Special Issue 1: 3–17.

Wareing, P. F. & al-Chalabi, T. (1985). Determination in plant cells. *Biologia Plantarum* 27: 241–8.

Wareing, P. F., Hanney, C. E. A. & Digby, J. (1964). The role of endogenous hormones in cambial activity and xylem differentiation. In *The Formation of Wood in Forest Trees*, ed. M. H. Zimmermann, pp. 323–44. New York: Academic Press.

Warren Wilson, J. (1978). The position of regenerating cambia: auxin/sucrose ratio and the gradient induction hypothesis. *Proceedings of the Royal Society of London*, B 203: 153–76.
Warren Wilson, J., Walker, E. S. & Warren Wilson, P. M. (1988). The role of basipetal auxin transport in the positional controls of abscission sites induced in *Impatiens sultani* stem explants. *Annals of Botany* 62: 487–95.
Warren Wilson, J. & Warren Wilson, P. M. (1961). The position of regenerating cambia – a new hypothesis. *The New Phytologist* 60: 63–73.
Warren Wilson, J. & Warren Wilson, P. M. (1981). The position of cambia regenerating in grafts of stems and abnormally-orientated petioles. *Annals of Botany* 47: 473–84.
Warren Wilson, J. & Warren Wilson, P. M. (1984). Control of tissue patterns in normal development and in regeneration. In *Positional Controls in Plant Development*, eds. P. W. Barlow & D. J. Carr, pp. 225–80. Cambridge: Cambridge University Press.
Warren Wilson, P. M., Warren Wilson, J. & Addicott, F. T. (1986). Induced abscission sites in internodal explants of *Impatiens sultani*: a new system for studying positional control. *Annals of Botany* 57: 511–30.
Watts, S., Rodriguez, J. L., Evans, S. E. & Davies, W. J. (1981). Root and shoot growth of plants treated with abscisic acid. *Annals of Botany* 47: 595–602.
Weiler, E. W. & Spanier, K. (1981). Phytohormones in the formation of crown gall tumors. *Planta* 153: 326–37.
Weisenseel, M. H., Dorn, A. & Jaffe, L. F. (1979). Natural H^+ currents transverse growing roots and root hairs of barley (*Hordeum vulgare* L.). *Plant Physiology* 64: 512–18.
Weiss, F. E. (1930). The problem of graft hybrids and chimeras. *Biological Reviews of the Cambridge Philosophical Society* 5: 231–71.
Weiss, P. (1950). Perspectives in the field of morphogenesis. *Quarterly Review of Biology* 25: 177–98.
Went, F. W. (1938). Specific factors other than auxin affecting growth and root formation. *Plant Physiology* 13: 55–80.
Went, F. W. (1941). Polarity of auxin transport in inverted *Tagetes* cuttings. *Botanical Gazette* 103: 386–90.
Went, F. W. & thimann, K. V. (1937). *Phytohormones*. New York: Macmillan.
Wenzler, H. & Meins, F. Jr (1986). Mapping regions of the maize leaf capable of proliferation in culture. *Protoplasma* 131: 103–5.
Wetmore, R. H. & Rier, J. P. (1963). Experimental induction of vascular tissues in callus of angiosperms. *American Journal of Botany* 50: 418–30.
Wheeler, A. W. (1971). Auxins and cytokinins exuded during formation of roots by detached primary leaves and stems of dwarf French bean (*Phaseolus vulgaris* L.). *Planta* 98: 128–35.
White, J. C. (1976). Correlative inhibition of lateral bud growth in *Phaseolus vulgaris* L. Effects of application of IAA to decapitated plants. *Annals of Botany* 40: 521–9.
Wickson, M. & Thimann, K. V. (1958). The antagonism of auxin and kinetin in apical dominance. *Physiologia Plantarum* 11: 62–74.
Wightman, F. & Thimann, K. V. (1980). Hormonal factors controlling the initiation and development of lateral roots. I. Sources of primordia-inducing substances in the primary root of pea seedlings. *Physiologia Plantarum* 49: 13–20.
Wightman, F., Schneider, E. A. & Thimann, K. V. (1980). Hormonal factors controlling the initiation and development of lateral roots. II. Effects of exogenous factors on lateral root formation in pea roots. *Physiologia Plantarum* 49: 304–14.

Wilcox, H. (1954). Primary organizations of active and dormant roots of noble fir, *Abies procera. American Journal of Botany* 41: 812–21.
Williams, M. H. & Green, P. B. (1988). Sequential scanning electron microscopy of a growing plant meristem. *Protoplasma* 147: 77–9.
Williams, R. F. (1974). *The Shoot Apex and Leaf Growth*. Cambridge: Cambridge University Press.
Winfree, A. T. (1974). Rotating chemical reactions. *Scientific American* 230 (6): 82–95.
Wolpert, L. (1969). Positional information and spatial pattern of cellular differentiation. *Journal of Theoretical Biology* 25: 1–17.
Wolpert, L. (1971). Positional information and pattern formation. *Current Topics in Developmental Biology* 6: 183–2.
Wolpert, L. (1981). Positional information and pattern formation. *Philosophical Transactions of the Royal Society of London*, B 295: 441–50.
Woolley, D. J. & Wareing, P. F. (1972). The role of roots, cytokinins and apical dominance in the control of lateral shoot form in *Solanum andigena. Planta* 105: 33–42.
Wright, M. (1981). Reversal of polarity of IAA transport in leaf sheath base of *Echinochloa colonum. Journal of Experimental Botany* 32: 159–69.
Wyndaele, R., Christiansen, J., Horseele, R., Rüdelsheim, P. & Onckelen, H. van (1988). Functional correlation between endogenous phytohormone levels and hormone autotrophy of transformed and habituated soybean cell lines. *Plant and Cell Physiology* 29: 1095–101.
Yeoman, M. M. (1984). Cellular recognition systems in grafting. *Encyclopedia of Plant Physiology*, New Series, 17: 453–72.
Young, B. S. (1954). The effect of leaf primordia on differentiation in the stem. *The New Phytologist* 53: 445–60.
Zajaczkowski, S., Wodzicki, T. J. & Bruinsma, J. (1983). A possible mechanism for whole-plant morphogenesis. *Physiologia Plantarum* 57: 306–10.
Zajaczkowski, S., Wodzicki, T. J. & Romberger, J. A. (1984). Auxin waves and plant morphogenesis. *Encyclopedia of Plant Physiology*, New Series, 10: 244–62.
Zimmerman, D. C. & Coudron, C. A. (1979). Identification of traumatin, a wound hormone, as 12 oxo-trans-10-dodecenoic acid. *Plant Physiology* 63: 536–41.
Zimmerman, R. H. (1972). Juvenility and flowering in woody plants: a review. *HortScience* 7: 447–55.
Zimmerman, R. H., Hackett, W. P. & Pharis, R. P. (1985). Hormonal aspects of phase change and precocious flowering. *Encyclopedia of Plant Physiology*, New Series, 11: 79–115.
Zimmermann, W., Woernie, D. & Warth, I. (1953). Genetische Untersuchungen an *Pulsatilla*. v. Die Entwicklung von Haaren und Spaltöffnungen bei *Pulsatilla*. *Zeitschrift für Botanik* 41: 227–46.
Zucconi, F. (1988). Epigenetic regulation in plants (minireview). *Israel Journal of Botany* 37: 131–44.

Author Index

Subject Index